AF710318

QA
159
E8
1928b

WITHDRAWN

A10202 671858

Columbia University

Contributions to Education

Teachers College Series

No. 324

AMS PRESS
NEW YORK

THE FUNDAMENTAL SKILLS OF ALGEBRA

BY

JOHN PHELPS EVERETT, PH.D.

PROFESSOR OF MATHEMATICS, WESTERN STATE TEACHERS COLLEGE
KALAMAZOO, MICHIGAN

TEACHERS COLLEGE, COLUMBIA UNIVERSITY
CONTRIBUTIONS TO EDUCATION, NO. 324

BUREAU OF PUBLICATIONS
Teachers College, Columbia University
NEW YORK CITY
1928

Library of Congress Cataloging in Publication Data

Everett, John Phelps, 1875-
The fundamental skills of algebra.

Reprint of the 1928 ed., issued in series: Teachers College, Columbia University. Contributions to education, no. 324.
Originally presented as the author's thesis, Columbia.
1. Algebra--Study and teaching. 2. Mathematical ability. I. Title. II. Series: Columbia University. Teachers College. Contributions to education, no. 324.
QA159.E8 1972 512.9'007'12 78-176759
ISBN 0-404-55324-9

Reprinted by Special Arrangement with Teachers College Press, New York, New York

From the edition of 1928, New York
First AMS edition published in 1972
Manufactured in the United States

AMS PRESS, INC.
NEW YORK, N. Y. 10003

ACKNOWLEDGMENTS

It is a pleasure to express a word of appreciation to those who have assisted in this study. To Professor W. D. Reeve, for having first envisaged the general idea of the problem, and whose foresight made the investigation possible by saving a great deal of original material that to an ordinary observer would have appeared to have already served its purpose. His scholarly human interest has been of great help throughout the undertaking. To Professor C. B. Upton, for incisive criticism and discriminating judgment affecting many items. To Professor P. M. Symonds, for suggesting desirable reorganization and amplification of the form which the work originally assumed. To my colleague, Mr. H. C. Christoferson, for many hours spent in friendly discussion which have clarified the content and expression of thought in many places. And finally to my wife, Nora Brewer Everett, whose assistance in collecting and systematizing the details of thousands of individual results and illustrations of pupils' actual accomplishments provided the material for the study for which I should probably have had neither the time nor the patience.

J. P. E.

PREFACE

This thesis represents the conclusions gained from an intensive and analytical study of written records of work done in algebra by several hundred pupils. The primary, and most important, source of information consisted of 285 exercises covering the entire course in ninth grade algebra and written on eighteen occasions throughout the school year, by 111 pupils in the University of Minnesota High School. The more than thirty thousand exercise-jobs of this particular study were taken as representing a fair sample of the type of work to be expected of high school pupils. A range of ability was represented extending from a low I. Q. of 96 to a high I. Q. of 158.

The conclusions were checked against the records made by 549 pupils, from 35 different schools, who wrote on 94 selected algebraic exercises just previous to taking the 1927 examination of the College Entrance Examination Board, and some illustrative material appearing in the sequel is taken from that source.

A further check was effected by using 45 selected exercises (which are given in the Appendix of this study) whose treatment requires a knowledge of all the algebraic topics (except problem solving) included in the 1927 Syllabus in Junior High School Mathematics by the State Department of Education of New York, and comparing the skills that have been listed with those which the exercises seemed to demand.

In a subject as old as algebra it would be undesirable to attempt to employ phraseology that is altogether new, and that has not been done. Many of the summaries of skills in their condensed form sound familiar. An investigation of the discussions preceding the summaries will reveal, however, that the basis upon which the skills have been determined is distinctly that of the exercise of thought processes which are naturally implied by the numerical relations which the content of algebra illustrates.

Early in the investigation, after the tabulation and diagnosis of the detailed records of many pupils covering every period and phase of first-year algebra, the opportunity was recognized for a predominantly statistical report of the type of materials and relations of manipulative skills included in first-year algebra. This could have been done with great profit and with relative ease. Inasmuch as implications of a statistical study are difficult for the ordinary reader to appreciate and apply, however, it seemed better to emphasize not a statistical summary, but applications to classroom procedure.

CONTENTS

THE FUNDAMENTAL SKILLS OF ALGEBRA

CHAPTER I

THE MEANING OF ALGEBRA

Mathematics has been defined by Benjamin Pierce as "the science which draws necessary conclusions." There are certain laws and relationships of algebra which obtain to such an extent that they are capable of being generalized and universally applied. Within well-defined limits there is in the laws and processes of algebra certainty of outcome. Within more arbitrary but fairly well-defined areas, there is substantial agreement as to what constitutes the topics, symbolism, and general forms of expression by which algebra is defined and understood. The fundamental aims of instruction are often vague at best. Even when objectives are clearly outlined, there is still great possibility of unsatisfactory mastery of the content material.

We need to obtain answers to the questions, Can the meaning of algebra be reduced to terms that will make its significance more apparent? What values has algebra to offer in terms of modern teachable units? and What are the psychological, as well as the logical, relations of these units to one another?

Algebra follows arithmetic in the school course, but it is much more than an extension of arithmetic. Algebra introduces ideas totally foreign to arithmetic. Directed numbers, for instance, and functional relationships are new to the person whose acquaintance with numbers extends no further than arithmetic. The apparently harmless statement that "algebra doubles the range of number" only feebly expresses the truth. Algebra not only introduces new numbers, it gives new meanings to old numbers; and these new meanings have to become significant in the mind of the pupil against the inertia of old associations that have seemed to be sufficient and complete. The pupil in algebra even has to unlearn some ideas that he has gained in arithmetic, such as the notion that a fixed value is attached to every expression of number value. The ideas of variables and of dependence may not be new to the pupil's experience, but they represent wholly new mathematical concepts in the transition from arithmetic to algebra. Likewise, the equation may have been used by the pupil to express a simple idea. Algebra introduces a new and different use of the equation, where it not only expresses an idea, but it also asks a question, where it stands not alone as the final form of expression of a fact, but as a part of an array of related statements and also of related *operations*.

Although the pupil encounters difficulties in the transition from arithmetic to algebra, the fact remains that he is well prepared by nature and ready for new mathematical concepts and a changed emphasis upon old ones. Where he has formerly been curious about things he now begins to do more reflective thinking; he is becoming increasingly anxious about values and is interested in weighing relative advantages in ways that have not engaged his attention before. In other words, he is impatient of authority. The pupil approaches the study of algebra at an age when his experiences, with separate objects and events have reached a climax in the sense that whatever realms of thought or space he may traverse, no other period of his life is likely to be so filled with sensations that are productive of new ideas or associations in themselves. He may continue to form new associations but it is extremely unlikely that they will be due to external stimuli to the same extent as in the earlier years of childhood. Rather, new ideas or new associations, if they do come, are likely to be influenced by the reorganization and modification of ideas and associations that have been acquired earlier. In the ninth grade the pupil's habits of thought have not become fixed, his mind stands upon the threshold of the world of concepts. Unconsciously, but often finally, he makes the choice which determines whether or not he shall be left to the monotony that eventually attaches to repetition of the same phenomena, no matter how thrilling they may be at first. In the high school the pupil is definitely taking on a philosophy of life which either restricts his mental vision or enables him to emerge from the naïve world of sensations into a new universe of thought where laws of quantitative measurement and organization of measurable forces give to events and things meanings never revealed by their mere existence. To observation of geography and science and language, and to all other experiences of the child, there may be added appreciation of laws not revealed by any one entity, but whose universal operation, once recognized, frees the mind from the burden of attempting to interpret life in terms of isolated detail, extends its control over all nature, and enables him to grasp some of the significance of a scheme greater than the apparent reality of things, and ruled by order and orderliness.

But all this organization of thought is of the mind's own creation, because its very process of statement of systematization puts into the world something not there before. Thus the earth becomes something more than the abode of physical and chemical reactions. With the measurement of forces and the reduction of their interplay to terms of simplicity through generalization, the social heritage of the ages, of which algebra is a part, enables one to find happiness not alone in variety of sensory stimuli but in the infinitely richer variety of ways in which the truths of mathematical laws emerge from the maze of ordinary, and often apparently confused, commonplaces of everyday existence.

The child needs algebra for what it will enable him to do, but infinitely more does he need it for the orderly way in which, in the midst of constant change, it will enable him to think.

It is one thing to appraise and classify incidents in terms of past experience. It requires a different type of ability to evaluate events and relations as a part of a universal scheme of organization, where laws reign throughout, but both economy and comprehensiveness of thinking require this ability, and it is the province of algebra to foster such an attainment.

In a cosmos where spiritual as well as physical facts rule, idealism often makes mistakes, but this does not mean that we should abandon our ideals. Rather we should seek to refine them so as to facilitate their attainment. This would seem to be especially true of algebra, which now, as in the past, continues to occupy the time and energy of a large proportion of pupils of high school age.

For more than a century algebra has had a prominent place in the secondary schools of the United States, and in the literature of this period there are many indications that the results of its study have not been satisfactory. Convincing evidence [1] has been

[1] W. D. Reeve, *A Diagnostic Study of the Teaching Problems in High School Mathematics*, p. 101. Ginn and Co., 1926.

accumulated showing that of the situations presented by the ordinary textbook course in algebra many pupils master a surprisingly small proportion of the facts and operations.

In the Appendix, pages 103 ff., there is presented a list of 45 exercises which were solved correctly by less than 50 per cent of approximately 1,200 ninth grade pupils in algebra throughout a school year, according to the study to which reference has been made. For convenience, five examples from this list are given here, with a statement as to the proportion of pupils who were able to solve them correctly.

TYPICAL EXERCISES[2]

Exercise No.	The Problem	Per Cent Right
3	Perform the indicated operations and simplify the result in the following: $5-\frac{2-3a}{5}-4a=$	15.8
7	Solve for r : $S=\frac{a}{1-r}$	21.4
10	$(2x^2\,m\,y^2)^2=$	22.0
12	Perform the following operation and simplify the result: $\frac{7a+2}{12a}-\frac{4a-3}{4a}=$	22.7
13	Solve for x : $\frac{5}{2-x}=\frac{3}{4+x}$	27.2

If an analysis is made of the manipulation skills necessary to solve the last problem, it appears that 91.9 per cent were able to perform the operation indicated by

$$9(-5x-2)$$

which is at least as difficult as performing the operation indicated by either $5(4+x)$ or $3(2-x)$; 97.3 per cent could find the sum of $3x+4x$, which is very similar to adding $3x$ to $5x$; 87.0 per cent could get the correct answer to the exercise:

$$\text{Add } +40,\ -10,\ -30,\ +18,\ -23,\ -31,\ +16,$$

which requires more sustained attention than any operation required in solving the given equation.

Taken separately it would appear as though there were cer-

[2] From Section I, Appendix of this study.

tainly no operation involving fundamental manipulative skills that could not be performed successfully by seven-eighths of all the pupils; yet the problem presented by the situation:

$$\text{Solve for } x\text{: } \frac{5}{2-x} = \frac{3}{4+x}$$

offers, in actual practice, such a wide variation from manipulative experience and ability as almost to justify the expression *totally different.*

It is evident that there is required for the solution of this equation some ability or abilities not contained in the so-called fundamental operations. What such abilities are, and their relations to one another, constitutes the main theme of this dissertation.

After occupying a prominent place in the public school curriculum for more than three generations there appears to be nothing particularly new, either in the extent to which algebra is taught or in the general attitude toward the subject.

A profound sense of its value, universal provision for its teaching, and dissatisfaction with results seem to have been prevalent since algebra was made a secondary subject. The catalogue of the University of Michigan for 1866-67, page 55, after listing among "particular [entrance] requirements for the classical course . . . algebra, through the first seven chapters of Ray's *Algebra,* Part II; or, what is equivalent, a thorough knowledge of the subject through Quadratic Equations, including the Calculus of Radicals," states further:

> Particular attention is called to the requirements in the pure Mathematics, for admission to the Freshman Class. So abundant are the facilities for preparation in these studies, and so essential are the requirements to success in the subsequent parts of the course, that the full measure of preparation will be demanded.
>
> Imperfect preparation in algebra is so common as to compel the conviction, that insufficient attention is given to this study in our Preparatory Schools, or that inadequate textbooks are used. . . . Nor is acquaintance with the processes of the science sufficient, without a thorough knowledge of the principles upon which these processes are based.

"Much water has run through the mill" since these words were written sixty years ago, but the statements sound modern. It is seldom that a comprehensive expression of fact, opinion, and advice can retain its cogency for so many years. Although criticisms directed at algebra have never ceased, neither gradual nor abrupt processes of administering it have served to change its external curriculum status in any marked particular.

Something of the extent to which pupils of high school age are affected at present by the content and methods of the algebra class is shown by a study made under the auspices of the North Central Association of Colleges and Secondary Schools in 1925.[3] In the nineteen states covered by the territory of the association, with 678,935 pupils, first-year English was the subject studied by the greatest number, but algebra stood next, with 975 enrolled for every 1,000 enrolled in first-year English.

Significant as are the revelations made by standard tests, and desirable as are both the mere ability to perform operations and the ability to comprehend facts, the most genuine contribution that algebra or any other school activity has to make is not in any kind of record of performances, but in the effect it produces upon the pupil. However wide may be the field for the direct employment of the results of algebraic processes themselves in situations that are clearly defined as algebraic, such accomplishments are small compared with giving to the pupil an ability to apply a more discriminating and a more competent exercise of judgment to his affairs generally.

Abilities in the application and organization of manipulative skills, rather than the mere skills themselves, seem to be most rare and most difficult to attain. In a recent address before the National Council of Teachers of Mathematics, a representative of the engineering department of the General Electric Company said that there is never any insurmountable difficulty in handling a mathematical formula once it has been obtained, but that it is exceedingly difficult to find a person who possesses sufficient insight into the meanings of mathematical processes to know when and how to apply them to problems as they arise. This statement may be taken as typical of the idea that mathematics, fraught with figures as it is, still feels the need of a kind of organization of ideas and the application of a congeries of thought processes that have not yet been reduced to a definite basis to such an extent that they can be readily translated into teaching practice.

[3] North Central Association of Colleges and Secondary Schools, *Proceedings, 1925*, p. 46.

Taking the results of able statistical inquiry that have been made, together with an analysis of subject matter, this study will achieve its purpose to the extent to which it succeeds in bringing to the attention of the reader suggestions for classroom procedure which are based upon relevant information that is available.

CHAPTER II

THE NATURE OF ALGEBRAIC SKILLS

Into the accomplishment of any piece of work to which it is in some measure difficult for almost any person, without previous training, to turn his hand, there enters an element generally designated by the word *skill*. Thus we speak of a skilled musician, or a skilled watch maker, or a skilled mathematician. Precisely what we mean by this term is obscure. It is comparatively easy to recognize the person who possesses skill; it is by no means easy to tell either what comprises the skill or how it may be imparted to another.

Because the word skill is generally applied to the ability to accomplish some result, the term is superficially connected solely with *fact* and *act*, and it must be admitted that frequently this application suffices; but certainly this is not true of the artist. No amount of scientific data with respect to color effect or proportion or the composition of paint or fabric is expected to develop an artist. Nor does any amount of study of sound waves produce an accomplished pianist—not even practice will do that. In the same way it seems difficult, if not impossible, to develop by any practicable amount of manipulative experience alone a comprehensive and satisfactory understanding of the real significance of algebra.

Long ago Thorndike demonstrated that a process by which a pupil performs an operation without some definite sense of its significance is not in itself educative, yet that is what constitutes a large part of our traditional algebraic instruction. A course in algebra often results in the direct opposite [1] of economy in the use of thought processes. Rather the subject seems to be made up of a great mass of isolated facts and operations. Many a pupil, after studying algebra, simply has that much more "on his mind" with the added responsibility of keeping it there and the inability

[1] C. H. Judd, *Psychology of Secondary Education,* p. 107. Ginn and Co., 1927.

to use the facts and operations that he apparently does know. Fortunately this is not true of all pupils, though it applies to altogether too many. The situation would be discouraging were it not for the fact that in some way a good many pupils do acquire facility in the use of algebra and the further fact that enough is known both of the nature of the learning process and the nature of the subject matter to offer abundant reason for the belief that much improvement would result from a more scientific adaptation to the mind of the learner of the materials of instruction, and especially from a more specific recognition and emphasis upon the fundamental values of algebra which include besides the manipulative skills, the fundamental skills of understanding. We seem reluctant to teach outright the meanings of processes. We seem to have a feeling that it is undesirable, or possibly unethical, to make perfectly plain just what it is all about. There still resides in the educational cult much of the love of mystery that surrounds the unlettered medicine man.

The fundamental processes of algebra are themselves social heritages, and as such, understanding of their nature and significance is dependent upon thoroughness of instruction. We do not wait for a pupil to acquire knowledge of addition by mere exposure to the possibility, or even the necessity, for addition. Why should we expect the pupil to acquire a real understanding of the more obscure meanings of the processes merely by exposure to the processes themselves?

Back of every act that belongs to the level of skilled occupations or professions there seems to belong some subtle factor of appreciation or a recognition of significance that is not put into operation by any clearly observable external stimuli and that is not exercised in accordance with, or accompanied by manipulative expressions that can be perceived. To this factor of meaningful significance as attached to an act we shall give the designation of *associative skill.* To acts which are self-contained in the sense that their performance itself indicates the essential nature of the mental processes which are involved, there will be applied the term manipulative skill.

All this is very different from saying that an associative skill cannot be detected, or even measured. Although it may not be observable in the act to which it is applied or in which it is exercised, its presence or absence, and possibly even its extent may be measured by tests which are wholly separate from the act itself. Thus one may be called upon to solve the quadratic equation

$$x^2 + 5x + 6 = 0.$$

It is simple to determine the method, that is, the manipulative procedure that is employed; but that act conveys no intimation whatever of whether one knows that x is a number, or that the results obtained have any significance in relation to the original equation. The answers to these questions are necessary, and indispensable, if the operation of solving the equation is to mean anything more than the performance of a more or less physical act, but the answers are not found in the mere act of solving the problem; nor is even the presence of a sense of awareness of the questions or answers found in the act.

The first step in an attempt to measure the skills of algebra which are of essentially a non-manipulative nature is to detect their presence as an essential part of the subject itself, and that is the object of this study.

Some research workers meet with an almost irresistible temptation to investigate only those characteristics of learning that are capable of statistical interpretation. This urge has revealed a great many facts dealing with the rate and amount of learning at different ages, the extent to which school activities correlate with actual life situations, and the extent to which learning under different designations, such as mathematics, science, language, and the life, does or does not embrace apparently identical elements of language or form or method.

To deny the benefit of the testing movement during the past twenty years would mean to disregard what is generally considered at the present time as the most outstanding educational development of the twentieth century. There is, however, more than a possibility that the testing movement has placed undue emphasis upon the easily measurable aspects of the educational processes, with consequent neglect of values which are not readily apparent; in other words, that the manipulative side of mathematical learning has been overstressed. This is no disparagement of the testing movement, because without it the inadequacy of manipulation as an end in itself might never have been so clearly revealed.

Even in so restricted a field as that of elementary formulas, ability to manipulate does not ensure successful results. In dealing with the Hotz algebra scales, more than one-half (54.4 per cent) of the pupils in the Proviso schools [2] made mistakes, not in the operations themselves, but in choosing the wrong operations.

In comparatively simple activities outside of mathematics expertness is frequently not so much a result of any observable or measurable act as it is a subtle appreciation of some fleeting characteristic of the process. Two blacksmiths may be given the same kind of welding job. The work of one may be uniformly smooth and finished in appearance, while that of the other is rough and inaccurate. No amount of observation—no test that may be given to the blacksmiths would reveal *why* the differences in the products appear, although the differences themselves are readily apparent. By scientific inspection and analysis of the material (in this case iron) that is being used, the precise reason for the inequality of the results may be revealed. In this instance the determining factor is the proper amount of heat to be applied; and this in turn is revealed by the color of the metal as it is subjected to the heat of the forge. Without instruction an intelligent and observant blacksmith will in time acquire the skill of striking the iron at the proper time, but he may attain this skill much more quickly without wasteful effort if its existence has been recognized and its value pointed out to him. Nearly everyone has observed that some cement sidewalks retain indefinitely a hard, smooth surface, while others very soon lose their surface finish and constantly give off particles of sand and dust as long as they are used. The difference between the sidewalks may be present when identically the same materials are used in identically the same proportions with every measurable factor of manipulation the same. As in welding iron, the correction comes not from a study of the operator who builds the sidewalk or from a study of anything that he does, but from a scientific scrutiny of the nature of the ingredients—sand, gravel, cement, water, which enter into the composition of the walk. Observation of processes in the construction of a sidewalk and tests of results reveal a great many things about economical operations of construction and the utility of the finished product,

[2] Edwin Schreiber, "A Study of the Factors of Success in First-Year Algebra," *The Mathematics Teacher*, Vol. 18, 1925, p. 154.

but the final determining factor which gives a lasting, durable, and thoroughly satisfactory finish requires the exercise of an ability not contained in any observable act of the operator. Whether or not the skill is employed, though, depends wholly upon the operator. The procedure requisite to secure the desired effect is simply that of applying the finishing tools at a particular stage in the chemical action that begins when the ingredients are put together. So far as the writer knows, there is no instrument or objective test that will determine when the proper stage has been reached; yet its recognition is so simple that most workers in cement are able effectively to apply the principle. Were there an instrument that would replace the associative skill of the operator, the essential conditions upon which the exercise of the skill would be based would still reside in the nature of the materials; and it should be observed further that the skill can be taught and that it is learned. Elementary as is the whole matter, literally years elapsed after cement construction became common before the ability was regularly exercised. For years the kind of a surface desired was clearly defined, but its attainment was a matter of accident or luck. The most painstaking care often failed; careless methods sometimes succeeded. Now the attainment of a single skill which is itself non-manipulative, and is only slightly, if at all, dependent upon practice, has replaced chance by certainty and has enabled the contractor to know that he can meet his specifications.

The essential superiority of accomplishment resides not in any act or in the use of special instruments or formulas, but in a sensitiveness to some element of the situation that results in the right action at the appropriate time. In algebra success often depends only in a secondary sense upon ability to perform a mathematical operation; the real test of efficiency lies in whether or not the mind has grasped some subtle understanding of relationships that sets the correct manipulative machinery in motion at the right time. The pupil who has learned that $a \times a$ is a^2, but who writes $x \times x$ as $2x$ may possess manipulative skill, but he certainly lacks associative skill in this particular situation. The pupil who reduces $\frac{16a^3 - 8a^5}{4a^2}$ to $4a - 8a^5$ shows that he possesses all of the manipulative skill necessary to handle the situation, but he has failed to exercise an associative skill of application. Likewise, the pupil who made a perfect record when confronted by such problems as:

Add $15a + 7a - 5a$
Solve $2x = 10$

but who was unable to deal successfully with the problem:

Solve $18x + 3x - 16x + 4x = 18$

demonstrated that he possessed all of the necessary manipulative skills that are needed, but that he did not have the essential associative skill.

The manipulative phase of algebra often receives attention to the exclusion of the associative skills. Manipulative skills are probably easier to teach than are the associative skills, but failure to incorporate the latter may account for the difficulty that many pupils experience in attempting to make desirable application of manipulative skill.

In algebra there seems to be at the present time scarcely any task which can be set that is sure to be treated with uniform success. Reeve in his *Diagnostic Study* [3] found it necessary to exclude only thirteen problems out of three hundred seventy-one because they were too easy, although all of the problems were very elementary. Thorndike [4] and Judd [5] both state, in effect, that algebraic formulas encountered in physics or chemistry constitute something quite different in the two situations.

All of this amounts to saying what is well known—that pupils do not acquire a fine sense of discrimination, or that they learn only a portion of what they might learn. Nor can we explain the failures of pupils in algebra on the basis of intelligence.[6] Although there is a high correlation between mathematical ability and intelligence, the correlation is by no means perfect and exceptions of a striking nature are frequent. In an intensive study of the accomplishment of 111 pupils in the University of Minnesota high schools, we find that a pupil with an I. Q. of 158 solved only 54.0 per cent of his problems correctly in tests extending through-

[3] Reeve, W. D., *A Diagnostic Study of the Teaching Problems in High School Mathematics*, p. 111. Ginn and Co., 1926.

[4] Thorndike, E. L. *Psychology of Algebra*, p. 74. The Macmillan Co., 1923.

[5] Judd, C. H. *Psychology of Secondary Education*, p. 140. Ginn and Co., 1927.

[6] Symonds, P. M., *Special Disability in Algebra*, p. 62. Teachers College, Columbia University, Contributions to Education, No. 132. 1923.

out the year, while another with an I. Q. of 109 solved 81.0 per cent of the same problems correctly.

Ten pupils with an I. Q. ranging from 104 to 116 and a median I. Q. of 108.5 solved 75 per cent of the problems correctly, while ten pupils with an I. Q. ranging from 118 to 158 and a median I. Q. of 121.5 solved 52 per cent of the same problems correctly.

In a study of *tendencies* the statements just made are, of course, without value, but in an inquiry made for the purpose of determining what the subject itself has to offer in the way of a challenge to the intelligence of the pupil, the study of individual cases and, especially, the reactions of individuals over a considerable period of time may go further to explain the reasons for discrepancies between apparent ability and actual accomplishment than could be gained by confining the study exclusively to aggregate results, or tendencies.[7] The high correlations found by several investigators between algebraic abilities and general intelligence causes these cases of striking departure from general tendencies to assume a significance that justifies inquiry into the reasons for such instances.

Coefficients of correlation and averages are highly significant in establishing the possibility for direct association between variables and for establishing means of achievement, but no coefficient of correlation is of value in determining the direction of functional relationship between two variables; that is, in establishing relations of cause and effect. These facts do not in the least detract from the value of, or the strict necessity for, statistical methods, but they do indicate a limitation that is inherent in mathematics itself and consequently in all places where mathematics is applied. It is worth repeating that the values that accrue from any mathematical work are not in the mathematics itself but in the manner in which the subject contributes to human conduct.

Psychologists would not be any more disposed to advocate the teaching of useless subjects than would any other teacher, but the laws of psychology apply equally well to the teaching of the timely and the obsolete, and to the good and the bad; just as the laws of physics work with utter impartiality either for or against human welfare.

In a study where manipulation of symbols might appear to be

[7] Symonds, *op. cit.*, p. 64.

the chief aim it is possible that the very minutiæ of detail, which would bore a bright mind that had already grasped all of the implications of a set of exercises, would make a strong appeal to a duller intellect, producing, in time, an indifference toward the subject on the part of the bright pupil and an adherence to the formality of stereotyped procedure on the part of the dull pupil that would give the latter a better record. Or it may happen that the bright pupil at first finds memorizing of forms so much easier than does the duller pupil, who may be forced to learn and comprehend laws and principles simply because he has not the time to memorize, that eventually the otherwise bright pupil is overwhelmed by too great a number and variety of separate forms and ideas to be carried in memory.

Because of the possibility for such situations as have been described, instruction which disregards, or considers as incidental, the associative skills, may be relatively more unfair to the bright pupil than to one who is duller. The study of individual behavior should contribute to the discovery of, causes of, and remedies for, failure.

The physicist and the builder of bridges both contribute to the construction of better bridges to the extent to which the lessons of the laboratory are incorporated in the actual work. Likewise, the psychologist and the teacher of mathematics, by wise and sympathetic collaboration, utilize together the discoveries of the psychologist's laboratory and improve the quality of instruction. So far as the utility of the bridge is concerned, getting the laws of physics incorporated into the structure is a most important necessity. So far as the pupil is concerned, getting the benefit of psychology into actual schoolroom operation is of more benefit to him than would be the discovery of still more benefits without their practical application. The person who harmonizes results of the theories and facts of scientific investigation with actual practice is himself making a reasonable contribution to education. Each type of contributor checks and reinforces the work of the other; both accomplish more than could be done were all the emphasis to be laid in one direction.

CHAPTER III

THE DETERMINATION OF SKILLS

Learning is a precarious activity, comparable only to teaching in the uncertainty of its result. We as teachers are told to *emphasize principles,* to *develop principles,* to look for transfer only where transfer is taught; to develop the power to generalize, but these statements are rather abstract declarations of general aims; they do not tell the teacher what to do to attain them.

All of these goals are reached eventually by a few individuals, but they are attained through the operation of methods that *expose* the pupil to the possibilities of algebra and yet do not place him in the way of acquiring algebra with either precision or certainty.

Of 1,204 pupils,[1] 97.5 per cent were able to solve correctly the first of the problems listed below; 89.3 per cent solved the second, 64.3 per cent the third and 53.7 per cent the fourth. These impor-

Problem	Per Cent of Responses Correct
(1) $4x + 5 = 17$	97.5
(2) $8x = 5x + 12$	89.3
(3) $\frac{y}{3} + 2 = 5$	64.3
(4) $\frac{z}{3} - 5 = 9$	53.7

tant facts reveal contrasts that are not explained by difficulties of manipulation, nor by differences in principle. The *principles* involved in (1) and (2) and those in (3) and (4) are identical, if it is considered that *more than* implies *less than* and vice versa. And it is quite evident that if it is ability to generalize or the appreciation of principles that was sought, the lessons immediately preceding the test from which these examples were taken

[1] Reeve, W. D., *Diagnostic Study of the Teaching Problems in High School Mathematics,* pp. 33, 104. Ginn and Co., 1926.

contained a large element of failure. The manipulative abilities of the four problems in order are:

In (1)
 Subtract 5 from 17
 Divide 12 by 4
In (2)
 Subtract $5x$ from $8x$
 Divide 12 by 3
In (3)
 Subtract 2 from 5
 Multiply 3 by 3
In (4)
 Add 5 to 9
 Multiply 14 by 3

If anything, the contrast in manipulative skills is more pronounced between problems (1) and (2) than between problems (3) and (4), since in the former case the subtraction of $5x$ from $8x$ might be considered as representing a more advanced type of accomplishment than the subtraction of 5 from 17. However, all of the pupils dealing with these problems were in the ninth grade; there is some presumption, therefore, in favor of their ability to perform these operations. Moreover, the fact that 97.7 per cent of the pupils were able to respond correctly to the direction, "Subtract $14x - 3x$," is fairly conclusive evidence that if the widest possible allowance is made for lack of manipulative skill, that circumstance still comes far from explaining the failure of so many pupils and still further does it fail to explain the range of accomplishment from 97.5 per cent to 53.7 per cent.

Without ability to perform the necessary algebraic operations, the solution of these equations would, of course, be impossible, but also the solution is impossible for a large number who possess the necessary manipulative abilities. Of the pupils taking the test in the ninth grade it is incredible to suppose that any considerable proportion, certainly not 35.7 per cent, were unable to perform the operations in (3), which consisted of subtracting 2 from 5, and of multiplying 3 by 3.

Evidently there enters into the solution of a simple equation some ability that is outside of, and distinct from, the ability to add, subtract, multiply, or divide. To abilities of this nature the name *associative skills* apply, and it will be the aim in succeeding chapters to point out some of the specific associative skills which are inherent in, and a necessary adjunct to, the understanding of algebra. The associative skills involved in the four problems above are: [2]

In all the problems:

x, y, and z are numbers.

In (1):

The second member, 17, is 5 more than $4x$.

(Therefore $4x$ must equal 12.)

12 is 4 times x.

(Therefore x must equal 3.)

In (2):

The first member, $8x$, is $5x$ more than 12.

(Therefore 12 must equal $3x$.)

12 is 3 times x.

(Therefore x must equal 4.)

In (3):

The second member, 5, is 2 more than $\frac{y}{3}$.

(Therefore $\frac{y}{3}$ must equal 3.)

3 is $\frac{1}{3}$ of y.

(Therefore y must equal 9.)

In (4):

The second member, 9, is 5 less than $\frac{z}{3}$.

(Therefore $\frac{z}{3}$ must equal 14.)

14 is $\frac{1}{3}$ of z.

(Therefore z must equal 42.)

Algebra Contrasted with Arithmetic

In their applications, either as methods of thinking or to physical situations, both algebra and arithmetic ask the same ques-

[2] For further detailed discussion of the associative skills employed see Association Skills of Interpretation, p. 59, and The Equation and Related Skills, p. 84.

tions, "When shall I perform a certain operation, and when I do perform it what shall I get?"

But the answers are radically different. To say that "algebra is generalized arithmetic" conveys just enough truth to make it dangerous. There is, to be sure, much of arithmetic in algebra, but the arithmetic of algebra is largely incidental; algebra has a field of its own, sharply distinct from both arithmetic and geometry.

When arithmetic asks its questions, "When shall I perform a certain operation, and when I do perform it what shall I get?" it expects only one kind of an answer and that answer is in the form of a definite number.

For example, arithmetic implies a question with respect to the multiplication of two numbers only in the presence of a specific situation. If I am interested in the amount of varnish necessary to finish off a floor I may find it necessary to multiply 26 by 14. The actual dimensions and the actual result for this particular instance are what I want. In algebra, on the other hand, the actual dimensions and the actual result get squarely in the way of what we want; namely, the general idea that *any* area of *any* rectangle is *lw*. Here there is actually no operation at all, but only a relationship expressed.

When algebra asks precisely the same questions as arithmetic it expects a totally different answer; namely, a certain relationship of number symbols; where these symbols are arithmetical, that relationship is definite; but where the measurements are *symbolic,* that relationship is indefinite in the arithmetical sense, but wholly definite in an algebraic sense.

To a degree algebra is more simple than arithmetic and its demands are more easily satisfied. Where arithmetic asks for the area of a circle whose radius is 4 and requires 4 to be multiplied by 4 and this product in turn by 3.1416, algebra asks for the area of a circle whose radius is r and is satisfied by writing the symbol for r multiplied by r in a certain relationship with π; πr^2 answers fully and satisfactorily our algebraic question.

The contrast in the two situations is not in the element of *abstractness,* but in the nature of what is demanded, being in the one case a number, and in the other a process with an indicated set of operations—a process with numerical implications. The symbolism of algebra by its very inclusiveness probably is less abstract than arithmetic. To the extent that algebra *implies* relations and results rather than states them explicitly, algebra is more abstract than arithmetic. When one thinks of eight trees, that act requires a tremendous amount of abstraction—far outstripping the incisive form of the expression. There is implied in the expression *8 trees* not only the idea of the number *8*, but in applying the number to the idea of *trees,* it is essential to forget or overlook a great mass of details, such as shape, size, peculiarities of bark and limbs, species of tree; and so on. At the same time the expression implies keeping in mind the general notions that distinguish trees from other objects, such as posts, shrubs, or fences. By comparison with *8 trees,* the expression *8a* might appear simple and even concrete, especially if, as is usually the case in algebra, *a* is considered as a *number.*

The same element of abstraction is present when we try to think of the price (or cost) of one dress if two are priced at $15. The two dresses may be of different color, different material, different style, different weight, different size even, the *quality* dress is abstracted from all of these by no means inconsiderable differences and for the time being that is all that may occupy the center of attention.

Likewise, the question, When shall I multiply one number by another? involves elements of abstraction, because I shall do this in situations of the same kind, where *kind* is wholly independent of the *things* involved. I shall multiply if I want to get the price of 5 pounds of sugar at 6.8 cents a pound. I shall do something totally different if I want to make a batch of candy. It isn't the things themselves that give me the cue as to what to do, it is what I want to do with these things. But I shall also multiply if I wish to determine the distance I can travel in 8 hours at 25 miles an hour. I shall likewise multiply if I wish to check my electric bill of 42 kilowatts at 7 cents a kilowatt.

Algebra does not require the pupil to become familiar with new ideas of abstraction as far as the things themselves or the symbols themselves are concerned. The especial feature of abstraction in algebra is found rather in the fact that algebra indicates processes and concerns itself with relations, whereas arithmetic is concerned with a single result applied to a particular question. If the postage charge for carrying a package a certain distance is 2 cents a pound, plus a flat charge of 6 cents, arithmetic is interested in getting the amount of the postage for a package of a given weight. Algebra does not require the weight to be given. Algebra is more liberal, less insistent upon the details of a particular transaction; but, on the other hand algebra is not satisfied until it has obtained an answer to the question regarding postage that covers *every* case of the same kind.

Arithmetic requires one to think through the situation thus:

The postage on this package, weighing 5 lb., is $5 \times 2¢ + 6¢$, and is not satisfied until the result has been reduced to its most simple form for this particular case.

Algebra requires one to think through the situation thus:

The postage on any package weighing x lb. is $x \times 2¢ + 6¢$, and is satisfied with this form, or it is satisfied to have any whole number substituted for x.

It is not so much *abstractness* that differentiates algebra from arithmetic as it is the element of *generality.*

You may wonder why the question in the opening paragraph of this discussion of Algebra contrasted with Arithmetic is stated in the form *When shall I perform a certain operation?* instead of *What operation shall I perform?*

There is very good reason for the distinction and a striking difference in the implications of the two questions. Algebra and arithmetic do not ask one to do many different acts, but they ask one to apply these acts to many different things and in many different circumstances. Elementary algebra and arithmetic require of the pupil that he shall be able to perform just four [3] manipulative operations—only four, *add, subtract, multiply,* and *divide. What operation shall I perform?* is, then, always one or more of just four—and these have all been learned and practiced in arithmetic. The operations themselves sound easy and they probably are easy; but algebra is not easy. The inconsistency between such a view of algebra and the actual difficulties that we know to exist in practice furnishes striking evidence of the necessity for considering algebra as something more than, or possibly different from, manipulation. The instances that have been cited on page 4 where there is very clear evidence that pupils possessed manipulative ability but were helpless in the

[3] The extraction of roots, or evolution as it is sometimes called, is often listed as a separate operation. Essentially, however, the extraction of roots, and all work with radicals, is an application of other fundamental operations rather than a distinct type.

presence of a situation demanding very elementary organization of the manipulative skills could easily be multiplied. It is when a choice of operations has to be made or an application is necessary that the greatest difficulties arise. Algebra is a subject that constantly requires one to make fine discriminations, to exercise judgment: it is a method of thinking quite as much as a collection of more narrowly technical abilities.

Probably the teaching of algebra tends toward overemphasis in two directions:

1. The "fundamental operations," instead of being fundamental in the sense that they can be learned as such and then applied, are more likely fundamental in the sense that they are essential throughout algebra; but they are best learned in connection with their applications. Fifty years ago it was customary to begin the study of algebra with addition, subtraction, multiplication, and division; but that is not the most common practice to-day.

2. We put too much teaching emphasis upon problems. In making such a statement it is easy to be misunderstood. Problems and problem solving measure power and furnish the most ready types of exercise in algebra. We could probably obtain the ability of problem solving better by actually solving fewer problems and discussing their implications more—by placing more emphasis upon the associative skills, and less upon variety of experiences.

The problem (No. 22, of the Appendix, page 104):

$$\text{Solve for } r\text{:} \quad C = \frac{E}{R + r}$$

which was actually solved by slightly more than one-third of 1,200 pupils, contains an element of difference as contrasted with $a = lw$ but its solution contains no principle not contained in the equations: $A = Pl_2$, where the value of l is to be obtained and $x + 5 = 9$, where the value of x is to be obtained; yet the last two were solved by 90.8 per cent, and 97.5 per cent of the pupils, while the value of r was obtained by only 36.9 per cent.

In these problems there was a fine opportunity for the teaching and the transfer of associative skills, but apparently this teaching was not done.

There are certain exercises which are sufficiently typical to be treated by themselves as:

$$\text{Solve: } x^2 + ax + b = 0$$
$$\text{Multiply: } (a + b)^2 \ldots$$

but this is not true of most problems, nor is it consistent with the aims of algebra to attempt to reduce problem solving to a series of pigeon-holed types.

As in geometry, the first work of the teacher is to see that the pupils are acquainted with figures, followed by an appreciation of relationships, so in algebra the first consideration is one of symbolism followed by relationships. The distinctive language of geometry is the figure, the distinctive language of algebra is the number symbol.

The symbol itself is capable of many different uses of a purely mechanical nature, but not until a genuine meaning attaches both to the symbol and to its relations to other symbols can a pupil be expected to deal with even apparent outward success with any situation outside those with which he has established direct contact. The pupil who obtains from $3a^2 - a^2$ the result 0 because $2 - 2$ is 0 is perfectly consistent if he has nothing more than manipulative ability upon which to rely. It is a fact that $2 - 2$ equals 0, and the pupil who reacts to the example in this manner has reached a conclusion that is wrong, not because it is irrational, but because he has failed to exercise an association skill of interpretation with respect to the meaning of the situation. As a matter of fact, the context often has more to do with determining a result than do the symbols that are employed. In the illustration above it happens that 0 is not the result, and cannot be under any interpretation, but it is also true that no other conclusion that might be drawn would be correct in the absence of further directive information. The pupil who has both the manipulative skill and the associative skill inherent in the idea that a^2 is a number is neither consistent nor so likely to be satisfied in getting the result 0. The pupil who possessed the associative skill would be obliged to resort to some operation that treated a^2 as the number capable of modification, not the 2 which constitutes the exponent. (See Associative Skills of Interpretation, p. 59.)

The Beginnings of Judgment in Algebra

Judgment is frequently evoked in the solution of problems and equations before a basis for its application has been established. In solving for l in the formula $A = lw$ there is, to be sure, plenty of opportunity for the use of judgment, though it is extremely difficult to detect whether it is exercise of judgment or something entirely different that has been employed in getting the "answer."

The pupil who writes $l = \frac{A}{w}$ may do so because he is keen enough to observe that such an *arrangement* of letters answers the demands. He may do so without any realization whatever that he is dealing with numbers or that any principles are involved. He differs from the pupil who writes $l = A - w$ or $l = \frac{w}{A}$ in that one gets a form that satisfies, while the other does not. In such cases one pupil is likely to get a mark of 100, the other a mark of 0, although neither has any real understanding of algebra; neither has used judgment; and both have attempted to do a piece of mental work of the level of grasping a door knob to open the door. Each has reacted to a mathematical situation only externally. Had he been rearranging furniture or books he would have engaged in the same kind of performance, but he would properly not have been suspected of performing a mathematical operation. Getting the wrong answer is interpreted as meaning that something is wrong either with the element of understanding or with the ability to perform operations. From the list of problems given on page 4 it will be observed that 77.3 per cent of the pupils were unable to perform the following operation and simplify the result:

$$\frac{7a + 2}{12a} - \frac{4a - 3}{4a}$$

although 93.9 per cent were able to perform the operations indicated by $3(5x - 1)$, which is as difficult as getting the result of multiplying $(4a - 3)$ by 3. Of these pupils, 72.4 per cent were able to subtract:

$$\begin{array}{r} 7(a + b) \\ -\underline{8(a + b)} \end{array}$$

which, again, is as difficult as the subtraction of $12a - 9$ from $7a + 2$.

For a large proportion of the pupils who correctly dealt with the last two problems there was evidently so little real understanding of the significance of the operations that their application in connection with fractions was attended with a mortality of 77.3 per cent. This again points to the conclusion that the element of interpretation of meanings is of as great importance as manipulative ability, and it further indicates that the ability to perform an operation demanding manipulative skill is slight evidence that the performance is necessarily attended by a sense of its significance. Significance is *associated* with applications of a skill, and in the case just cited there enters both the associative skills of the fundamental operations themselves and the associative skills of the new notion introduced by the fraction. The associative skills of fractions are discussed in detail on pages 66-69.

It is possible for the very simplicity of algebra to prove its undoing, at least in the process of learning. Generalization means of necessity that there shall be comprehended in a symbolism a summary of a large number of meanings. To give every specific idea its own form of expression would necessarily leave no opportunity for generalization. It follows, therefore, that the *language* itself of algebra is not characterized by the fine distinctions that govern ordinary spoken and written discourse, where a letter, a word, or a phrase is seldom expected to convey a complete idea. Rather in the ordinary usage of language these elements are combined into a more or less complicated sentence which is modified in many ways to fit into particular situations. This is true to such an extent that the simple sentence, "The boy studies," is capable of being expanded into a variety of sentences which employ every modification known to grammar; but, conversely, unless it is so expanded its meaning is confined to a narrow range of significance.

Contrast the sentence in quotation marks above with the formula

$$a = lw$$

which is among the most simple that can be found in algebra. With no modification of form whatever this *expression* may have the following meanings (and usually only one is needed at a time):

1. Something, called "a" equals something, called "lw."
2. Some number, a, equals some number lw.

3. Some number, a, equals some number, l, multiplied by some number w.

4. The area of a rectangle equals the product of the length and width. (A mere statement of fact.)

5. To get the area of a rectangle you *multiply* the length by the width.

6. a is a function of l and w.

7. An arrangement of "letters" with no numerical meaning whatever.

8. a is so related to l that a change in the value of l produces a change in the value of a.

9. a is so related to w that a change in the value of w produces a change in the value of a.

Suppose that a pupil sees in the expression $a = lw$ only the significance indicated by interpretation No. 7 above. This may hinder him not at all in "solving" for l. Solving may mean simply writing $l = \frac{a}{w}$.

So far as this particular operation is concerned that is the answer, and no more need be asked or expected. But this cannot be considered as algebra merely because letters are employed. Presently this same pupil may reach the physics class and be absolutely confounded when confronted with the privilege of solving for v in $k = pv$. If the pupil has learned to solve for l in the formula $a = lw$ in a formal way without reference to fundamental meanings there is little to transfer to the new situation embraced in: Solve for v in $k = pv$. To the extent to which definite associations have been formed with $a = lw$, or with mere manipulation involving its solution, there is reason to expect that there is no generalized notion in the pupil's mind and therefore nothing to enable transfer to take place. Contrast a pupil who possesses manipulative dexterity in solving for l in the formula $a = lw$ with the pupil who has acquired the simple associative skills of seeing in lw the product of l and w and the additional associative skill of recognizing in the equation a statement of the fact that a is l times w. The one has nothing better than an imitative ability to fall back upon, the other has a deep sense of relationships that goes back to the fundamental notions of number meanings.

There has been a great deal of loose thinking concerning this question of transfer. Transfer requires something to transfer as well as ability to transfer. In the case cited the differences lie far deeper than in the names of the classes or a change of rooms and instructors. The two situations have next to nothing in common for the student who has merely learned to get a result in algebra. For such a student the degree of similarity probably extends about as far as that between a page printed in English and one printed in Latin, where the same type and punctuation are employed. The pupil who has been taught in algebra that results of a manipulative routine are sufficient to gain high marks has been deceived if he thinks that this will give him an ability that can carry over into physics.

The associative skill which accompanies the solutions of $a = lw$ for l and $pv = k$ for v goes further than manipulation and further than ability to generalize as to symbols. The associative skill would perceive not only that a, l, w, p, v, and k are numbers, but that, in the first instance a is w times the number that is wanted; and in the second instance k is p times the number wanted.

CHAPTER IV

LEARNING IN ALGEBRA

As pointed out in previous chapters, the associative skills give meanings to operations and relations. Another, and quite as important, advantage of associative skills lies in the possible economy of thought, which their recognition and use affords. In learning it is a well-known fact that a mass of unrelated details presents to the learner a problem very different from that offered by the same number of details when related to some unifying idea.

The development of scientific testing in algebra has brought into clear relief two significant conditions. These are the wide divergence between aims and accomplishment,[1] and the prodigious number of facts and operations [2] to be mastered. The different abilities suggested by the usual course in algebra run up into the thousands. A summary [3] of what is expected in arithmetic totals 1,680 combinations, to be acquired over an interval of six or eight years.

For every one of the arithmetical combinations there is one in algebra, but this is only the beginning of the purely numerical relations of algebra. Algebra not only makes use of all the *numbers* of arithmetic and immediately doubles their range by the introduction of the idea of directed number; it admits any symbol or any combination of symbols that may be thought of. To be sure, the literal symbols usually employed are conventionalized so as to make their number comparatively small, but even so, the mass of number expressions grows rapidly. If only ten literal symbols were to be employed in algebra and only Arabic numerals below 10 included in the original data, no symbol being used more than once and not more than four literal symbols used

[1] Reeve, W. D., *A Diagnostic Study of the Teaching Problems in High School Mathematics.* Ginn and Co., 1926.

[2] McKeown, Kate, "Learning in First-Year Algebra." An unpublished Master's thesis in the library of the University of Oklahoma.

[3] Osburn, W. J., *Corrective Arithmetic,* p. 11. Houghton Mifflin Co., 1924.

with any numerical coefficient, there would be rendered available of mere number expressions the stupendous total of 117,705 numbers. This array of numbers would arise merely from using each of the digits with the combinations that would arise by taking the ten letters four at a time. These would not be comparable with an equal number of Arabic numbers because of the way in which algebraic numbers are constituted. (See Associative Skills of Interpretation, page 59.)

In this array there exist no laws of place and intrinsic values. The absence of place values renders the interpretation of algebraic numbers more difficult, since in the number *abc* there resides also the equal values of *acb, bac, bca, cab,* and *cba.* In arithmetic any three-figured number, say 267, means just that. If 267 is *the* number that is applicable to a certain situation, there it stays. No one later is ever called upon to recognize or use 672 as exactly equivalent to 267 merely because the same separate symbols are employed.

We will exclude, for the present, intrinsic values, because we are discussing the intricacies of only the literal notation. It is obvious, however, that were we to ascribe to our limited ten letters no value greater than 9 in any case, we should immediately reach a group of number expressions of whose extent it is difficult for the mind to form any comprehensive idea. With thousands of expressions capable of assuming all of the relative positions taken by numbers in equations, fractions, sums, and the like, it becomes evident that if specific forms are to be mastered as such, the time devoted to algebra will permit of consideration of a very limited proportion of the total. But even if it were possible to acquire familiarity with forms of expression in an individual and detailed manner, that would still leave unsatisfied the demand for a cultivation of the ability to generalize. What may be called the physical limitations of algebraic number expression and one of the main purposes of the study itself are both opposed to a course in algebra which operates as an organization of habits pertaining to specific arrangements of algebraic symbols.

The essence of algebra, however, is not mere number *expression,* but number *relations,* and this term includes considerably more than it does in arithmetic. Whereas in the latter a specific application of any relation, say of multiplication, exists between two numbers, as 6×7, or a specific relation like that of cost and number exists in arithmetic in a given specific situation, in algebra both of these types of relationship lose all semblance of application exclusively to a particular situation. The expression $a \times b$ contains only the relation of multiplication. Its advantages are great, since the very fact of its applicability to any situation, whether of cost and number, volume and number, length and number, and so on, renders the arrangement available for use anywhere; but this very fact places on the learner a burden not imposed by arithmetic. It is not impossible that the very habits that the pupil has acquired, through years of acquaintance with arithmetic, of ascribing a particular value to every number stands much in the way of getting easily and quickly a comprehension and appreciation of generality, which is the very essence of algebra.

What is Learning in Algebra?

A comparison of achievement between two pupils should take into account fluctuations of accomplishment in the same individual. The expression "one hundred per cent mastery" stands in need of more precise definition than it seems to have received. If the phrase means to a given stimulus always the right response under all circumstances, then it might better be eliminated from our language completely, because it is difficult to get even a machine to respond always in the same manner. The development of the nervous organism has affected the area of certainty of outcome in inverse ratio to the capacity for self-determination.

Another meaning for "one hundred per cent mastery" might be a degree of attainment sufficient to satisfy the demands of the situation in which the ability is to be employed. This would have the disadvantage of varying arbitrarily and would be of little use in school practice because of the uncertainty of the kind of position that the pupil would eventually fill.

A still better meaning to attach to "one hundred per cent mastery" would be "that degree of mastery which can be attained under measurable conditions." This definition also would vary, but under control. It would have the advantage of making it possible to set up in advance a norm of achievement suitable to the variables entering into a unit of work. The law of diminishing returns might be given a chance to operate. Thus a pupil of a certain I. Q., after a certain amount of practice in adding directed numbers, might be certified as having reached proficiency of a certain desirable level.

It is doubtful, though, if the expression "one hundred per cent mastery" can ever be used satisfactorily. *One hundred per cent* is so closely associated with the idea of all there is to anything that it is likely to stand for too much where human accomplishment is involved. A person seldom reaches "one hundred per cent mastery" in terms of any measure of an activity unless a mistake carries its own penalty and the penalty is immediately apparent. The very abstractness of mathematical procedure precludes the possibility of awareness of error, except after some interval of time. In checking a result, for instance, the discovery of a wrong answer may produce annoyance and the resolution not to commit the same blunder again, but even then the connection between the act and the result is very remote as compared with striking a wrong note on the piano or taking the wrong step on a flight of stairs.

The expression "optimum mastery" would be an easier one to which to attach meaning because it would be possible to bring it into harmony with the facts as they are. Ability to attach the correct sign to the product of two directed numbers is fundamental to nearly all work in algebra, but is it of sufficient importance to require that a pupil should be able to handle directed numbers with unerring skill before being allowed to work with problems and other operations in which this particular skill functions? The desirable procedure, for instance, would seem to be to allow the skill, "correct sign" to be acquired [4] gradually in applications which both directly and indirectly but certainly penalize the mistakes. But the penalty for errors should embrace the idea of self-conviction of error as well as the idea of *wrong answer*.

In the paragraph above we have discussed the advantage of recognizing an optimum mastery from the standpoint of practice, but there are other equally strong arguments against dwelling on any topic with the expectation of always obtaining the correct response to a stimulus. The degree of attainment in any skill that a pupil evinces at a given time may be quite out of harmony with what he has done before or will be able to do in

[4] Thorndike, E. L., *Educational Psychology*, Vol. I, p. 134. Teachers College, Columbia University, Bureau of Publications, 1913.

the immediate future. The incorrect response does not certainly denote inability and, for many algebraic situations, the correct response is not necessarily an indication of the ability that is supposed to be represented.

Thorndike has pointed out and Symonds [5] has elaborated into specific illustrations seven cases where a pupil, from the standpoint of the extent of his learning, *could* make the correct response but is prevented from doing so by mental set or external conditions. This implies that degree of mastery is always a question of the control of a considerable number of factors which in themselves have no especial connection with the particular piece of work to be accomplished.

A pupil called upon to factor $a^2 - b^2$ may give the incorrect response because he does not know the correct response, but he also may fail because between the impact of the stimulus and the recording of the result there intervenes some external distraction. A steam calliope underneath the window may suddenly start up, with a train of suggestions of activities far removed from algebra. The extent to which the ordinary pupil can be taught to concentrate to the exclusion of unusual sensory stimuli is limited. These factors must of necessity be taken into account in any attempt to reach conclusions affecting the meaning of mastery. The best accomplishment of any pupil over a somewhat wide range of time might be taken as measuring what this optimum is, provided it were possible to determine whether the effort is actually the *best*.

In a study of the accomplishments in algebra of 111 pupils in the high school of the University of Minnesota during the academic years 1922-23 and 1923-24, an examination was made of the record of these pupils in 285 exercises or problems written on eighteen occasions at approximately two-week intervals throughout their ninth school year. Because problems and work with coefficients as used in addition and subtraction are readily identifiable, there were selected from the 285 exercises 28 in which work with coefficients predominated. The extent to which the initial accomplishment was reflected in subsequent work was investigated by examining the record made by each individual. The initial test included the following exercises.

[5] Symonds, P. M., in *The Mathematics Teacher,* Vol. 15, p. 93, 1922.

Add the terms as indicated in problems (1) and (2):

(1) $3x + 4x = .$
(2) $3x + 20x + 17x + 5x =$
(3) Subtract as indicated: $14x - 3x =$

Perform the indicated operations in problems (4) to (6):

(4) $15a + 7a - a =$
(5) $10x - 2x + 8 - 3 =$
(6) $15y - y + 4y =$

Of the 111 pupils taking this test, 28 made a perfect score. In subsequent tests which contained fractional coefficients, given on five different occasions and including twenty-two problems, the best record was 100 per cent, made by one pupil whose I.Q. was 132. The poorest record was 36.3 per cent, made by a pupil whose I.Q. was 121. It is worth noting that in the simple accomplishment under observation, 3 pupils with I.Q.'s higher than that of the pupil with the best record, one rating 139 and two rating 135, made records on the problems of 77.2 per cent, 77.7 per cent, and 40.9 per cent, respectively.

Although only one pupil out of 111 was able to make a perfect score on both the original learning lesson and the subsequent application of lessons extending over a period of a year, as compared with 110 pupils who failed in one or both respects, it was noted that of the 83 pupils who failed, as shown by the test, in their first attempt to learn the addition of coefficients, eight of them in the subsequent applicational tests made records as good as, or better than, any of the pupils whom the first test recorded as perfect. Apparently keeping these eight pupils at work with coefficients until they could make a perfect score would not have been attended with effects more beneficial than those of the amount of practice that they had received, in spite of the fact that their understanding of coefficients was deficient, as revealed by the first test.

Ability to get the right answer to a problem like:

$$\begin{array}{rr} \text{Add} & 2\,a \\ & \underline{3\,a} \end{array}$$

may be due to memory only, or it may be due to imitation. An attempt to *generalize* by the introduction of many literal symbols is a favorite device. If a pupil can respond correctly when a is replaced by b or x or by ab or xy, or any other simple or complicated number, we are ordinarily satisfied. But here again, memory, not understanding, may play the chief part. Memory of the preeise problem may be replaced by memory of the kind that gets the result by saying, "Add the Hindu-Arabic numbers algebraically; do nothing to the rest of the expressions, except write *another* like them after the sum." The worst thing about this "method" is that it works—it is perfect from that aspect, but such a pure manipulative ability seems to lack nearly everything that is required when the pupil comes to a situation which demands some organization of the material previous to the act of manipulation instead of the mere performance of some algebraic operation.

The associative skills of interpretation (see p. 59) and of addition-substraction provide a check which takes care of the process of addition-subtraction in those particulars in which something more is demanded of the pupils than the mere act of arithmetical combinations of coefficients.

In the illustrative problem in subtraction (Appendix, p. 107), with its variations of wrong results, the difference obtained by subtracting $7b$ from $-12b$, as obtained, is just everything possible that could be gotten from any addition-subtraction combination of the coefficients; being $-5b$, $-19b$, $+19b$, and $+5b$. The 26 varieties of results provide exactly as many wrong as right responses, thirteen of the answers being the correct $-19b$, and the other thirteen shared by $-5b$, $19b$, and $5b$. Apparently these pupils were much more impressed with the necessity for getting something done than with the attaching of any significance to their acts.

A commentary very much to the point concerning the inadequacy of manipulative skills and the failure of teaching to produce satisfactory results, because of imperfect treatment of such situations, is found on page 198 of Smith and Reeve's *The Teaching of Junior High School Mathematics* (Ginn and Co., 1927), where the authors say,

> As an illustration of the failure to master certain of the fundamental skills in algebra a recent study shows that 74.4 per cent of a group of over

600 ninth grade pupils could simplify the expression $7x + (-3x)$, and that 70.9 per cent could perform the following subtraction:

$$\begin{array}{r} +8a - 12b - 156 \\ -5a + 7b - 75 \\ \hline \end{array}$$

In other words, the former was nearly as difficult as the latter. The same study shows that for the same group of pupils it was considerably more difficult to simplify $4x - 6y - 5x - 3y$ than to perform the subtraction

$$\begin{array}{r} -ab + a^2 - 3b^2 \\ ab - a^2 + 2b^2 \\ \hline \end{array}$$

even though the latter involved the subtraction of $+2b^2$ from $-3b^2$, which should be as difficult as anything in the former. In the first case 29.3 per cent of the answers were correct, whereas in the second case the score was 65.3 per cent.

There are doubtless various possible explanations of the facts cited above. That there should be such a difference in the scores shows how marked had been the failure to analyze objectives and to drill adequately upon the constituent skills involved.

Here is very strong evidence to support the belief that among these pupils the manipulative skill which they had acquired in dealing with one situation was attended with so slight associative skill of understanding that the external form of the problem constituted a much more potent influence in determining results than did any genuine comprehension of the nature of the algebraic operations which the exercises involved.

It might take a pupil longer to acquire apparent facility in subtraction by emphasizing the associative skills of the operation as related to the number scale, than by teaching the simple rule of "change sign and add," but the acquisition of these skills would certainly provide at least the opportunity for the exercise of a set of satisfiers and annoyers that manipulation alone does not include.

CHAPTER V

DIRECTED NUMBERS

Meaning of "Greater Than" and "Less Than"

(integers)

In arithmetic the idea of comparative size of two numbers, the notion of *greater than* and of *less than* goes along with the numbers. The number 8 is greater than 5; 15 is less than 21. These are fundamental properties of numbers where size begins with 0. The effect of the introduction of directed numbers is not alone to change the scale of numbers, but the most important thing it does is to give to the numerals themselves implications of size very different from those which arithmetic has always depended upon to answer the question as to which is the greater. For instance, in algebra 8 remains greater than 5, but —8 is *less* than 5; 8 is greater than —5, but —8 is less than —5. This is confusing to the person who reckons number values as plus and minus from 0; it is very plain to the person who recognizes each of these numbers as belonging to a number scale and as occupying a definite relative position with respect to every other number of the scale.

The kind of comparisons made above between 8 and 5 and between 15 and 21 would also be confusing were it not for the fact that to the numbers 8, 5, 15, 21 there attaches a vivid significance of relative size. Whether care has been taken or not to impress upon the pupil the fact that 6 is one more than 5, all his experience suggests and reinforces such an idea with respect to these two numbers and all other numbers, and in many ways these facts check one's work. Addition of two numbers in arithmetic always yields more than either of the numbers; subtraction always yields less than the greater, or minuend.

In algebra *more* ceases to be a correlative of addition and *less* no longer is a part of the idea of subtraction.

In arithmetic the process of counting almost inevitably conveys a sense of the relative values of the numbers. Probably no one ever counts with algebraic or directed numbers. The result of this is that while experience of an intuitive or practical nature does not tend to contribute largely to a pupil's sense of appreciation of the significance of directed or algebraic numbers, at the same time the arithmetical experience that the pupil has had tends to handicap him severely in getting a comprehension of the meaning and value of numbers when 0 is no longer used as the place of references from which values are to be reckoned. These circumstances necessitate an especial emphasis upon associative skills designed to give to the pupil a sense of meaning and an understanding of the appropriateness of directed numbers in those situations to which they apply.

Algebraic integers may be arranged in order, as may arithmetical integers, but in algebra 0 does not stand at the end of the scale; there is no least number of the algebraic scale. Suppose that we arrange the numbers, for convenience, as increasing from left to right according to the following diagram, where the divisions of the line are employed simply to give an illustration of what the numbers might represent,

—6 —5 —4 —3 —2 —1 0 1 2 3 4 5 6

As ever (in arithmetic) 3 comes before and is less than 4; 4 comes before and is less than 5. That part is easy. It is not quite so apparent that —4 comes before and is less than —3; and that —5 comes before and is less than —4; but we can see that fact if we imagine that we are starting at some point off at the left (this may be any point whatever). If we count successively from left to right we shall be constantly *increasing* the distance from our starting point. Now it is evident that the distance from our starting point is greater when we are at —3 than it is when we are at —4. In other words on the serial number scale —3 is greater than —4, and so on for all of the negative numbers. All of the numbers behave in the same way, with the greater number *always* on the right.

In this algebraic or directed scale of numbers 0 retains a position of importance, but it is no longer the starting point for the measurement of the significance of the numerals of the scale. In point of significance of value the minus symbols at the left of 0 take whatever significance they have from their position relative to all of the other numerals on the scale, not, as commonly supposed, because of their relation to 0. For instance, the value denoted by — 3 would be exactly the same were the symbol to be replaced by φ without any negative sign. In point of measurement of values, the only significance of the minus sign as applied to numbers at the left of 0 is to identify them. So far as algebraic operations of addition and subtraction upon them are concerned there is no property whatever possessed by a number at the right of 0 that is not possessed by a number at the left of 0, and vice versa, although nearly all of the rules of manipulation imply the belief that — 5 and + 5 differ because one is minus while the other is plus.

To speak of a negative number as though it were different from a positive number because of its quality is like using the word *transpose* in dealing with the equation. Treating a negative number as though it contrasted with a positive number in any different sense whatever from the contrast between two positive numbers is to ascribe to the negative number a mythical property that it does not possess.

Any positively directed number possesses a certain element of measure, but so also does any negatively directed number. Further, any so-called positive number may be negatively directed and any so-called negative number may be positively directed with respect to other numbers. The numbers that are labeled with a minus sign, as — 3, — 25, and the like, take this designation simply to place them in the scale, not because the sign as thus used has significance in any operation of addition and subtraction with such numbers.

The essential difference, as to algebraic meaning, is not one of minusness and plusness, but a difference of value, + 5 standing for 10 more than — 5; — 5 standing for 10 less than + 5; but this is precisely the distinction that exists between 9 and 19, or between — 2 and 8, or between — 82 and —72. All the distinctions are of values in the number scale, as the numbers *increase in value from the left;* not in the fact that the minus numbers lie on one side of zero and the positive numbers on the other. The fact must be kept in mind that this discussion has to do strictly with the values expressed by the numerals, including 0, in the directed number scale. As a measure of absolute value,

0 assumes properties which are not possessed by any other number.

The important thing that the invention of positive and negative numbers did to numerical measurement was not to introduce a new measure of values *beginning* with 0, but to introduce a new measure of values beginning anywhere we please. Only the place where a number belongs, not its value, is conveniently reckoned from 0.

One of the most confusing, and also one of the most unscientific, of algebraic notions is tied up with an attempt to reconcile the number scale with statistical measures and physical events, without reference to the real meaning of the directed character of the numbers used. For instance, one of the commonest illustrations of the meaning of positive and negative numbers, as encountered in the classroom, is that of actually having some money as contrasted with being in debt. "The amount had" being designated by a + sign and "the amount in debt" by a — sign. As a matter of fact, the measure of the two kinds of values is in terms of the same kind of numbers. The + and — signs, or the plus and minus numbers, are not in the least needed to express the ideas of "having" and "not having"; the words take care of those ideas.

An illustration of directed numbers that goes no further than to call "amount possessed" by the name *plus* and "amount owed" by the name *minus* is wrong and misleading in that for such ideas directed numbers were never needed and for such situations probably never would have been invented. The primary notions of measure are those of *greater* and *less*, and number scales operate accordingly. It is perfectly possible for given measures to designate exactly opposite tendencies in the measurement of different magnitudes, but the number scale itself would be intractable and unreliable if at any arbitrary or capricious point it were suddenly to change its meaning and somewhere begin to mean *more* where before it had meant *less*. The number scale does not behave that way, as it would have to were the usefulness or the meaning of directed numbers to be applied to the measurement of "having" and "not having."

The use and application of the idea of directed numbers appear only in the measurement of such quantities as exist in a sense not measurable by one type of number alone. In the case of the illustration partially completed above, precisely such a condition of inability to measure by means of numbers of one direction is found in recording the individual's *financial status.* That element of financial status is a quantity which retains the qualities of more and less throughout its entire extent. It could not be expressed or measured by either plus or negative numbers alone, and as to size, it begins at one end of the scale and progressively increases *all the way* toward the other end.

The invention of the expression *directed number* is probably one of the most useful terms that has been introduced into mathematics. Every number in the scale is positively or negatively directed with respect to every other number; 9 is positively directed with respect to 7; —8 is positively directed with respect to —11; 6 is negatively directed with respect to 15; and —17 is negatively directed with respect to —14. The quality of positiveness extends over the entire scale, as does also the quality of negativeness. The name "minus" number will probably always be retained, but it should be recognized for what it is, a mere name, when applied to a number, just as 6 is a name different from 8.

The symbol + (or —) now assumes *three* instead of the orthodox two meanings: (1) a sign of direction, (2) a sign of operation, (3) a mark of identification, usually called *quality.*

A unit of value in the scale of directed numbers means precisely what it would were numbers to be considered as not having direction. The designation of a number of units without reference to direction is in terms of what is called *absolute* number. Thus, if we are talking about the number scale in terms of the illustration given on page 37, the figure 3 itself refers to a position in the scale, but 3 *units* refers to a distance comprehended by this measure anywhere on the scale, where the distance is thought of without reference to direction or, perhaps better still, where the distance is thought of as though direction, as applied to numbers, did not exist.

All this requires one to distinguish between the meaning of + (or —) as applied to an expression of value in the number scale and the same sign as applied to a number used as an operator. In the addition of — 6 and — 6, for example, these two numbers are identical, but their rôle in addition is by no means the same. One of these numbers is added to the other, they are not added to each other, except in the sense that either may be taken as the number to which the other is added. In arithmetic this distinction between number-operated-from and number-operating-on is of slight consequence, because no one ever seems to think that he is adding different somethings because the numbers to be added are different, but in algebra the expressions commonly encountered in textbooks give the impression that if $+6$ and -3 are to be added the $+6$ is a plus number and the -3 is a minus number. This plays the mischief, because the most fundamental notion of addition is that of a change in some quantity of a measurable property produced by either an identical or a different quantity of the same measurable property. How, then, may we speak in algebra of the addition of a positive quantity to a negative quantity? An attempt to justify such a question immediately involves us in difficulties, which can only be avoided by running into another difficulty of changing our definition. But it is unnecessary to make any change in definition if we will relieve the operating number of quality entirely. This is perfectly consistent, because we already know that the unit of measure behaves the same way; that is, it is the same whether toward the minus or toward the plus end of the scale. We may call the number to be added *plus* (or *minus*), if we please, but we should understand that this word *plus* (or *minus*) gives us a direction of addition, not a property of the number. The quantity to be added is precisely the same regardless of the sign that stands before the number.

The important conclusion is that all operations with algebraic numbers, so far as numerical implications are concerned, are with *absolute* values of the numbers. That is we do not add $+3$ or multiply by $+3$; what we do is to add 3 or multiply by 3. In the same way we do not add -5 or multiply by -5; we add 5 or multiply by 5, and in both cases the signs tell us the *direction* that the operation takes. Only the absolute values of numbers may ever serve as operators.

Summary of Associative Skills of Directed Number

1. The number scale contains both the positive and negative numbers, but the scale does *not* begin at 0 and *run in opposite* directions.

2. *Less* lies toward one end of the scale, *more* toward the

other, but the scale has no beginning. It is conventional to think of *less* as belonging to the left-hand end of the scale and *more* as belonging to the right-hand end. In other words, the numbers of the scale increase from left to right.

3. Positively directed or negatively directed numbers cover the entire scale. With respect to any given number of the scale, every other number that lies at the left of the given number is negatively directed; and every number that lies at the right of the given number is positively directed.

4. The sign + (or —) has three entirely distinct meanings in algebra: the first meaning denoting something about the place that the number occupies in the scale of numbers; the second denoting direction on the scale of number; and the third denoting an operation of addition (or subtraction).

5. Number is used in algebra in two senses: non-directed and directed. The non-directed or absolute value of a number is its significance as applied to a number of units, without reference to direction. The directed value of a number is its absolute value modified by the element of direction.

6. It is the absolute value of number that is always used as an operator.

A LESSON ON THE MEANING OF DIRECTION AS APPLIED TO NUMBERS

P —9 —8 —7 —6 —5 —4 —3 —2 —1 0 1 2 3 4 5 5 7 8 9

Suppose that we are measuring the distance from some point *P* off at the left.

On the number scale, which of the two numbers 2 and —6 represent the greater distance from *P*?

In this sense which of these two numbers is the greater?

Of the two numbers 0 and —1, which is the greater?

Which is the greater of the two numbers —10 and —4?

Which is the greater of the numbers + 10 and + 4?

Which is the lesser of each of the pairs given above?

Have you reached any conclusion as to which is the greater of two numbers generally? [Yes, the number on the right.]

Which is always the lesser of two numbers? [The one on the left.]

In comparing a negative and a positive number, how much do you need to know of the numbers in order to decide which is the greater [or lesser]?

In comparing two negative (or positive) numbers how much do you need to have given about the numbers in order to determine which is the greater? Or the lesser?

What are the numbers that are less than + 4? [+ 3, + 2, and all other numbers at the left of these?]

What are the numbers that are less than —4? [—5, —6, and all numbers at the left of these?]

What are the numbers that are greater than —4? [—3, —2, and all numbers at the right of these?]

Are there any numbers in the scale that are greater than 0? That are less than 0?

0 (zero) is greater than what numbers? Less than what numbers?

With respect to any given number, all other numbers that are less than the given number are said to be *negatively directed* with respect to the given number; all other numbers that are greater than the given number are said to be *positively directed* with respect to the given number.

Name a number that is negatively directed with respect to + 5. With respect to — 3.

Name a number that is positively directed with respect to — 7. With respect to 24.

Can you make a statement that will include all of the numbers that are positively directed with respect to + 7? — 8? 0?

Make a statement that will include all numbers that are negatively directed with respect to + 9; — 4; 0.

Is the first of the following pairs of numbers positively or negatively directed with respect to the second: 8 and 2, — 6 and — 4, 3 and 5, 4 and — 9, — 6 and 7?

If two numbers are of unlike sign, how much must you know of the numbers in order to determine whether the first is positively or negatively directed with respect to the second? [Their signs.]

If two numbers are of the same sign, what must you know of the numbers in order to determine whether the first is positively or negatively directed with respect to the second? [Their names.]

In any given situation involving positive and negative numbers, how many of the numbers might be positively directed with respect to another number? [Any number from *one* up to one less than the total number of numbers in the situation.]

Associative Skills of Algebraic Addition and Subtraction

As demonstrated in the previous section, the understanding of algebraic addition demands an element of interpretation not included in arithmetic, because algebraic addition greatly extends the meaning of the operation.

Fundamentally, of course, any operation derives its scientific meaning from a definition, but not from a rule. Psychological meanings are more likely to be derived from experience with applications of an operation, but not necessarily if the applications are made in a formal manner. A rule for an algebraic operation, say of addition, may be applied because it works, but not because it *is* addition.

In algebra the laws of arithmetic, and of the universe, appear to be dethroned, because we may take the numbers designated by 3 and 4 and by adding them we get 7, 1, — 1, and — 7, according as 3 and 4 take the signs + or —. "If the signs are alike, add the numbers and prefix the common sign. If the signs are unlike, subtract the numerically smaller from the numerically greater and prefix the sign of 'the greater' " is a rule that works, but it works just like a key to the adding machine, mysteriously, without rhyme or reason.

In arithmetic the operation indicated by the word *add,* possesses, first, an element of simplicity and, second, an element of automatism that with algebra in the first instance certainly, and in the second probably, is seldom attained. Addition in arithmetic means beginning with a number where it is and going in a certain direction by a definite measured amount. Direction as used here is with reference to the number scale. To be sure, we do not often think of the operation in that way, because the number combinations that have been learned in addition equip us with a sort of seven-league-boots ability to stride over intervening obstacles and arrive all at once at a destination quite far removed from the starting point, but these facts do not eliminate the starting point nor change the essential meaning of the operation.

Addition in algebra also means beginning with a number where it is and going in a certain direction by a definite measured amount, but whereas, in arithmetic, the direction is always the same, in algebra the direction is one of two.

In arithmetical subtraction the operation is performed by beginning with a number at a point in the number scale and proceeding in a certain direction by a definite measured amount. Again, as in addition, the direction in arithmetic denoted by the minus sign is always the same; while in algebra the direction is one of two. In arithmetical subtraction the operation is limited to the case where the number "taken away" is less than the number "taken from"; else a number not in the scale is obtained. In algebra no such limitation is enforced; there is no such thing any more as an integer that is not in the scale. All this necessitates for algebra the learning outright of a [1] large number of new addition and subtraction combinations; otherwise the imposition of new definitions or new concepts of the extended meaning of numbers. As shown on page 29 it is practically impossible to learn the algebraic combinations as the arithmetical combinations are supposed to be learned. An associative skill which might offer an economical and sufficient foundation for extending the facts already learned in arithmetic to cover the directed numbers of algebra should not only lessen the time necessary to acquire facility in the use of such numbers, but at the same time increase the precision and certainty of the pupil's work.

The meaning of algebraic addition is that the number to which another number is to be added simply indicates the point of departure for the addition: the number to be added indicates by its sign the direction of the addition, and by its absolute value the measure of the addition. Thus in the scale below, if + 3 is to be added to + 5, + 5 indicates the point of departure on the scale, 3 indicates the measure of addition and the + of the 3 indicates the direction of the addition. Evidently if one starts at

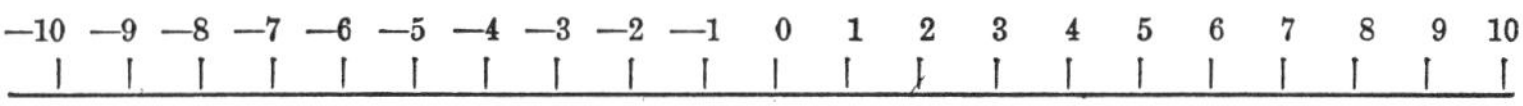

+ 5 and measures three units toward the right, the result is + 8.

If instead of adding + 3 to + 5, we wish to add — 3 to + 5, the starting point remains the same. Now since the sign of the number to be added is —, we *add* by going to the left (in a minus direction), 3 units and the result is 2.

In adding + 7 to — 3 the point of departure is — 3, the direction is +, and the sum obviously + 4; and so on.

A rule or a stereotyped method of procedure in addition will undoubtedly speed the operation itself, once its meaning has been acquired, but the rule lacks both the element of meaning and the basis for providing a check of the validity of the results. As a piece of abstract manipulation it seems often to be a matter of indifference as to whether the sum of — 5 and + 4 is — 9, + 9, — 1, or + 1. The fact that pupils seldom, in arithmetical addition, make mistakes that are utterly inconsistent with the numbers added in that the result is less than either, gives some reason for expecting that the introduction into algebraic addition of a definite, tangible basis for checking the process might narrow the margin of error which ordinarily attends the operations.

[1] Smith, D. E., and Reeve, W. D., *The Teaching of Junior High School Mathematics*, p. 202. Ginn and Co., 1927.

It should be observed that addition is performed in terms of the absolute values of numbers. The sign of the operator in every instance where a mathematical operation of any kind is performed possesses the sole function of indicating the direction of the operation.

Summary of Associative Skills of Addition

1. The sign of the number to be added to another always indicates the direction of the addition on the number scale.
2. The number to which a number is to be added indicates the point of departure for the addition. In this number the *sign* is only a mark of identification.

Manipulative Skills of Addition

The illustrative tables (I and II) of manipulative skills are given mainly for the purpose of furnishing a graphic picture of the practically prohibitive number of such skills from the standpoint of the establishment of a response to such situations on a habit plane gained by practice in each of the particular situations.

If the order of operation is from the top down, thus adding the upper to the lower, the illustrative situations may be obtained by interchanging the words "lesser" and "greater" in columns one and two, and "Number" and "0" in columns four and five.

In Table II the manipulative skills of addition with algebraic numbers having integral coefficients are illustrated. The term x may be thought of as representing any algebraic number whatever, no matter how complicated.

As in the case of Table I, if it is desired to add from the top down, the illustrations already given in Table II may be made to serve if appropriate changes are made in the column headings.

In both tables the column rather than the line arrangement of the numbers has been employed for convenience. It is recognized that probably a majority of algebraic situations involving two numbers are of the line variety, but the column arrangement is also common, and the principle of procedure is in no wise affected by passing from one to the other.

In Table I there are presented 45 jobs each requiring its own interpretation. Table II increases the number of jobs requiring an interpretation by 87 more, making a total of 132. The actual number of expressions that may occur in each instance is unlimited. Suppose one considers each specific skill contained here as belonging to a group of 9 others. That immediately gives a total of 1,320, or a sufficient number to keep a pupil busy for some time were he to depend upon his memory for them. The inclusion of decimals would nearly double this number of combinations.

Table I gives illustrations of manipulative skills of addition with integers only:

TABLE I*

THIRD MEANING OF + EXPRESSED BY THE IDEA OF ADDITION

	LOWER NUMBER ADDED TO UPPER					
	Lesser Added to Greater†	Greater Added to Lesser	Two Numbers Equal	Number Added to 0	0 Added to Number	0 Added to 0
Signs of Both Numbers Expressed						
Plus added to plus	+7 +3	+2 +9	+6 +6	0 +7	+3 0	0 0
Plus added to minus	−8 +2	−3 +5	−7 +7		−5 0	
Minus added to plus	+6 −3	+7 −9	+8 −8	0 −2		
Minus added to minus	−4 −3	−2 −5	−3 −3			
Signs of Only Minus Numbers Expressed						
Plus added to plus	7 3	2 9	6 6	0 6	3 0	
Plus added to minus	−8 2	−3 5	−7 7			
Minus added to plus	6 −0	7 −9	8 −8	0 −2		

[*Table continued on page 48*]

* The style of this table is suggested by Smith and Reeve's *The Teaching of Junior High School Mathematics*, p. 200. Ginn and Co., 1927.

† The words "lesser" and "greater" are here used, not with reference to the algebraic scale, but with reference to their absolute values.

TABLE I (*continued*)

THIRD MEANING OF + EXPRESSED BY THE SIGN						
	LOWER NUMBER ADDED TO UPPER					
	Lesser Added to Greater †	Greater Added to Lesser	Two Numbers Equal	Number Added to 0	0 Added to Number	0 Added to 0
Signs of Both Numbers Expressed						
Plus added to plus	+7 +(+3)	+2 +(+9)	+6 +(+6)	0 +(+7)	+3 +(0)	
Plus added to minus	−8 +(+2)	−3 +(+5)	−7 +(+7)			
Minus added to plus	6 +(−3)	7 +(−9)	8 +(−8)	0 +(−2)		
Signs of Only Minus Numbers Expressed	7 +3	2 +9	6 +6	0 +7	3 +0	

† The words "lesser" and "greater" are here used, not with reference to the algebraic scale, but with reference to their absolute values.

Even a slight amount of practice is practically impossible on all of the combinations that are easily possible. It would seem as though it were fairly apparent that limitations of time practically preclude the possibility of giving a pupil adequate mastery of even the facts of addition, except through the acquisition and operation of an associative skill which serves to reduce several hundred manipulative skills of algebraic addition to a number little, if any, greater than the manipulative addition skills of arithmetic.

In order to perform algebraic addition there is needed besides the associative skills of directed number, page 41, the specific addition associative skills, included in the idea of more and less involved in the number scale. The *addition* of a plus number begins at the point in the scale indicated by the number "added to" and proceeds toward the right. The *addition* of a negative number begins at the point in the scale indicated by the number "added to" and proceeds to the left.

These last two statements are mere rules to the person who sees in addition nothing but a manipulative process. The state-

TABLE II

THIRD MEANING OF + EXPRESSED BY THE IDEA OF ADDITION									
	LOWER NUMBER ADDED TO UPPER								
	Lesser Added to Greater*	Greater Added to Lesser*	Numbers Equal	Coefficient of Upper Number Understood	Coefficient of Lower Number Understood	Coefficient of Both Numbers Understood	Upper Number 0	Lower Number 0	Coefficient of Sum Understood
Signs of Both Numbers Expressed									
Plus added to plus..........	$+7x$ $+3x$	$+2x$ $+9x$	$+6x$ $+6x$	$+x$ $+3x$	$+2x$ $+x$	$+x$ $+x$	0 $+2x$	$+9x$ 0	$+x$ 0
Plus added to minus.........	$-8x$ $+2x$	$-3x$ $+5x$	$-6x$ $+6x$	$-x$ $+3x$	$-5x$ $+x$	$-x$ $+x$		$-9x$ -0	$-8x$ $+7x$
Minus added to plus.........	$+6x$ $-3x$	$+7x$ $-9x$	$+8x$ $-8x$	$+x$ $-3x$	$+5x$ $-x$	$+x$ $-x$	0 $-2x$		$+4x$ $-5x$
Minus added to minus.......	$-4x$ $-3x$	$-2x$ $-5x$	$-3x$ $-3x$	$-x$ $-5x$	$-7x$ $-x$	$-x$ $-x$			
Signs of Only Minus Numbers Expressed									
Plus added to plus..........	$7x$ $3x$	$2x$ $9x$	$6x$ $6x$	x $3x$	$2x$ x	x x	0 $2x$	$9x$ 0	x 0
Plus added to minus.........	$-8x$ $2x$	$-3x$ $5x$	$-6x$ $6x$	$-x$ $3x$	$-5x$ x	$-x$ x		$-9x$ 0	$-8x$ $9x$
Minus added to plus.........	$6x$ $-3x$	$7x$ $-9x$	$8x$ $-8x$	x $-3x$	$5x$ $-x$	x $-x$			$4x$ $-5x$
THIRD MEANING OF + EXPRESSED BY THE SIGN +									
Signs of Both Numbers Expressed									
Plus added to plus..........	$+7x$ $+(+3x)$	$+2x$ $+(+9x)$	$+6x$ $+(+6x)$	$+x$ $+(+3x)$	$+2x$ $+(+x)$	$+x$ $+(+x)$	0 $+(+2x)$	$+9x$ $+0$	$+x$ $+0$
Plus added to minus.........	$-8x$ $+(+2x)$	$-3x$ $+(5x)$	$-6x$ $+(+6x)$	$-x$ $+(+3x)$	$-5x$ $+(+x)$	$-x$ $+(+x)$			$-8x$ $9x)$
Minus added to minus.......	$-4x$ $+(-4x)$	$-2x$ $+(-5x)$	$-3x$ $+(-3x)$	$-x$ $+(-5x)$	$-7x$ $+(-x)$	$-x$ $+(-x)$		$-4x$ $+(-5x)$	

* See note to Table I.

ments are an expression of a generalized conclusion to one who possesses besides the definition of + and — as directions the associative skill of connecting the ideas of *more* and *less* with the number scale.

Algebraic Skills of Subtraction

Subtraction occupies a peculiar position among the fundamental operations. Nothing is accomplished by algebraic subtracting that is not achieved by algebraic adding. If either addition or subtraction as distinct names and processes could be eliminated from vocabulary and thought nothing in the way of mathematical facility or flexibility would be lost and the subject of mathematics would be materially improved by simplification.

To a person in the eleventh century who multiplied by "doubling" and "halving" those were necessary processes. We get along to-day not only without the processes, but also without the words themselves, so far as their having any technical operative sense is concerned. It is not impossible that the same lapse into oblivion is destined eventually for the word *subtraction.* Subtraction is found as a distinct topic in all of our textbooks, and most people believe in it as they believe in anything whose existence and utility they have never questioned.

If pupils are to be taught subtraction it would be doubtful psychology to begin their instruction with a pronouncement of its uselessness or a statement to the effect that they will be able to do nothing, after learning subtraction as a separate operation, that they could not already do by addition. The teacher, however, should know the facts in order to avoid the confusion that often accompanies an attempt to reconcile the operations of addition and subtraction.

We have already seen, page 45, that the addition of a positive number simply has the effect of yielding a result indicated by a number on the scale at the right of the point from which the addition proceeds, while the addition of a negative number has the effect of yielding a result indicated by a number at the left of the point on the scale from which the addition proceeds. Subtraction as ordinarily interpreted is the reverse of addition; the reason that this statement is meaningless, in effect, is that addition already takes *both* of the directions of algebraic directed numbers.

From this point subtraction will be treated as though it were actually a separate process.

Algebraic subtraction of two numbers may be thought of as the process of finding what number must be added to one number to produce another. The number added is called a *difference*, but there are some language difficulties to be overcome in dealing with the operation. The expression

$$\text{Take away } -3 \text{ from } +4$$

is perfectly definite; it means, "Find the number which, added to -3, gives a sum $+4$." The number scale associative skills of page 45 make it apparent that the number to be added is $+7$. Or if the problem is stated in the similar form, but with different numbers, as *take away* -9 from -7 the same associative skill gives the answer $+2$. But if the problem in subtraction is stated as *what is the difference between* -3 and $+4$, it is necessary to give the expression an arbitrary or technical significance which the language itself does not convey, because if -3 is taken away from $+4$ the result is 7, but if $+4$ is taken away from -3 the result is -7.

In view of the ambiguity of the phrase *difference between* it would seem wise to use some other expression unless the context were such as to indicate clearly which number is to be taken away. Once the problem has been reduced to writing its interpretation is well standardized. The number expression

$$2-(+3)$$

means that 3 is the number to be taken away. The meaning of subtraction being the finding of the *number which must be added to the one taken away to produce the other*, the result of taking away $+3$ from $+2$ is -1. Likewise, the number expression

$$\begin{array}{r} 7 \\ -9 \end{array}$$

where it is understood that the two numbers are to be subtracted, always means that the lower number is to be taken away from the upper. In these two particulars, form of expression and interpretation are well standardized. The number which must be added to -9 to produce 7 is, according to the associative skill of addition (see page 45), the number 16.

It will be observed that the process of subtraction is one of starting from the place in the number scale denoted by the number to be taken away and going in the direction which must be taken to obtain the other number given. The relative positions of the two numbers on the number scale thus always give both the directed sign and the absolute value of the result. This associative skill of reconciling the meaning of the operation with the number scale implications furnishes at once both a precise idea of the *direction* of the result and a rough indication of the absolute measure of the result.

The fact that opportunities for comparison between addition and subtraction are so numerous introduces the possibility for a confused state of mind toward the two operations to such an extent that it may be better to develop the associative skill of subtraction analytically rather than graphically.

Let the relations between remainder, minuend, and subtrahend in algebra be represented by

$$d = m - s$$

(see The Equation and Related Skills, page 84).

This is reasonable, since subtraction is the process of taking one number from another, and the result of this operation is a difference.

Now if m takes for the time being any fixed value whatever, it becomes apparent that any increase in the value of s diminishes d and any decrease in the value of s increases d. These facts may not be intuitional. Their algebraic counterpart needs further explanation. The ideas contained in an interpretation of the changes in the value of d affected by changes in the relative values of m and s are explained in the discussion of associative skills of relation, p. 76, and they may be illustrated inductively by the following scheme of substitution:

Suppose that in the formula above, m assumes the fixed value 9; then

If s is 4, d is 5
If s is 6, d is 3
If s is 8, d is 1
If s is 10, d is -1
If s is 12, d is -3
If s is 14, d is -5

and so on, as may be readily verified from the scale of numbers, page 45.

Let us investigate what happens when s decreases.

Likewise by referring to the scale of numbers, we may see that:

If s is 2, d is 7
If s is 1, d is 8
If s is 0, d is 9
If s is -1, d is 10
If s is -2, d is 11
If s is -3, d is 12

The facts of the last three subtractions become apparent if we note that each decrease by 1 in s increases the value of r by 1. The subtraction of a negative number thus gives the same result as the addition of a positive number and the subtraction of a positive number gives the same result as the addition of a negative number, which is consistent with the statements made on page 50 concerning addition and subtraction.

It may be noted that this interpretation of subtraction actually parallels the process of subtraction. The explanation sometimes encountered where the difference between 9 and -3 is shown by actual count on the scale to be 12 is not a process illustrative of subtraction, but of addition.

For an explanation of the process of subtraction when the minuend is negative, let m be any negative number and apply the same set of values of s.

Summary of Associative Skills of Subtraction

Subtraction means adding to the number *taken away* a number sufficient to produce the other, and both direction and absolute value of the result are determined by the relative positions of the two numbers on the number scale.[2]

Algebraic Skills of Multiplication

Multiplication, like other operations, must exhibit a continuity throughout the number scale. The rule of signs, "Like signs give plus, unlike signs give minus," in multiplication has

[2] An illustration of all of the types of manipulative skills of subtraction is to be found in Smith and Reeve's *Teaching of Junior High School Mathematics*, page 200. Ginn & Co., 1927.

the advantage of brevity and it works, but it gets squarely in the way of addition and subtraction and, at the best, conveys no idea whatever of the mathematical relationships between the numbers of the algebraic scale which make the rule operative.

Four distinct types of multiplication with directed numbers are to be observed, which may be illustrated by allowing a to represent the multiplicand and b to represent the multiplier.

$$\begin{array}{ll} (1) & +a \times (+b) \\ (2) & +a \times (-b) \\ (3) & -a \times (+b) \\ (4) & -a \times (-b) \end{array}$$

In all these types a, the quantity to be multiplied, may be considered as a number to be added or subtracted, beginning from 0, while b may be considered as denoting whether a (+ or —) is to be added or subtracted, and how many times. When b takes the sign + it means that a is to be added, when b takes the sign — it means that a is to be subtracted. This is a primary, fundamental meaning of the symbolism as it is employed, and the meaning must be obtained in the same way that the vocabulary of a foreign language is acquired—through a direct statement of the significance of the forms as they are used.

Once the facts have been stated that $a \times (+b)$ means that a is to be added b times, from 0 as the initial point, and that $a \times (-b)$ is an expression for *a multiplied by* $-b$, and that this means *a subtracted b times,* the interpretation of what this amounts to is a matter for the exercise of the associative skills of subtraction.

In the same way the expression $-a \times (+b)$ has a primary meaning of $-a$ added b times. Also the expression $-a \times (-b)$ has a primary meaning of $-a$ subtracted b times.

In the first case if a is considered as a unit on the number scale, the addition of $+a$ any number of times is going to give $a+$ sum (see Associative Skill of Addition, page 43) and the absolute value of this sum is going to be a taken b times. A more abbreviated way of writing this result is $+ab$ or ab.

In the second case the expression $+a \times (-b)$ means that $+a$ is to be subtracted and that the number of times it is to be subtracted is b. The subtraction of $+a$ once is going to yield a difference of $-a$. That is, the number that must be added to

$+a$ to produce 0 is $-a$. (See Associative Skill of Subtraction, page 53.) The subtraction again of $+a$ from $-a$ is going to yield $-2a$, and so on. The continued subtraction of $+a$ any number of times is going to yield a $-$ difference, and the absolute value of this difference is going to be b times a. An abbreviated way of expressing this fact is $-ab$.

In the third case $-a \times (+b)$ means that $-a$ is to be *added* b times. The addition of $-a$ once is going to yield $-a$; the addition of $-a$ twice yields $-2a$, and so on. The sum in this case is going to be minus and the absolute value is going to be represented by a taken whatever the *number of times* is, in this case b. The abbreviated form in which the value of a minus number, whose absolute value is a times b, is written is $-ab$.

In the fourth case the expression $-a \times (-b)$ means that $-a$ is to be subtracted and that the number of times it is to be subtracted is b. The subtraction of $-a$ once yields a difference of a; the subtraction of $-a$ from this difference in turn yields $2a$, and so on. The continued subtraction of $-a$ until it has been subtracted b times, yields a difference whose sign is $+$ and whose absolute value is ab. This may be written $+ab$ or ab.

A somewhat detailed suggestion for introducing the idea of directed numbers in multiplication will be found on the following page. The more concrete exposition found there may be clearer than the generalized discussion of meanings as given here.

From the standpoint of immediate results it cannot be doubted that the acquisition of the associative skills of multiplication will consume more time than will be required to learn the rule "like signs give plus, unlike signs give minus," but one method involves the element of understanding, the other requires the pupil to respond in definite ways, but blindly, to certain situations. If the main object in algebra is to enable pupils to think in terms of processes, it would seem that consideration should be given to making these processes take on a significance that extends beyond their mechanical aspects.

Summary of Associative Skills of Multiplication

1. The sign of the multiplier indicates whether the process of multiplication is to be thought of in terms of addition or of sub-traction.

2. As soon as the nature of the process (that is, whet

addition or subtraction) has been determined, the associative skills of addition and subtraction determine the sign of the result.

3. The sign of the result in multiplication is + when a plus number is added $[(+a) \times (+b)]$ or when a minus number is subtracted $[(-a) \times (-b)]$.

4. The sign of the result in multiplication is minus when a plus number is subtracted $[(+a) \times (-b)]$ or when a minus number is added $[(-a) \times (+b)]$.

A TYPICAL CLASS EXERCISE

There follow suggestions for typical exercises in introducing the associative skills of directed number in multiplication:

DIRECTED NUMBERS IN MULTIPLICATION

A lesson in the multiplication of directed numbers presupposes skill in addition and subtraction. A first lesson might employ the number scale directly, but very soon the scale as a visible aid should be discarded.

—8	—7	—6	—5	—4	—3	—2	—1	0	1	2	3	4	5	6	7	8
|	|	|	|	|	|	|	|	|	|	|	|	|	|	|	|	|

If 4 is added to 0 what is the sum?

If another 4 is added to the sum above what is the result?

This sum represents the addition of how many 4's?

Might the sum 8 be thought of as having been obtained in any other manner than as the result of $4 + 4$? [Yes, as $4 \times (+2)$.]

If another 4 is added to the last sum above, what is the sum now obtained?

This sum represents the addition of how many 4's?

Might we think of this sum as having been obtained in any other way than as $4 + 4 + 4$? [Yes, as three times 4; which is written $4 \times (+3)$.]

What is the result of adding -5 to 0?

If to the sum -5 above we add another -5, what is the final sum?

Of how many -5's does -10 represent the sum?

Could -10 be thought of as having been obtained in any other manner than by adding (-5) and (-5)? [Yes, (-5) and (-5) is equal to $-5 \times (+2)$.]

This type of exercise may be continued until the pupil is able to generalize in terms of $-7 \times (+3)$, $-5x\,(+8)$, and $-n \times (+m)$, where $-7 \times (+3)$ is read "minus seven multiplied by plus three." This is a better form than "minus seven times plus three," because the latter form of expression gives no intimation of which number is multiplied and which is the multiplier. As to an understanding of the process, this distinction is essential.

What is the result of adding —5 to 0?

How many —5's does this represent the addition of?

If to the sum —5 above we add another —5, what is the final sum?

How many —5's does this represent the sum of?

Could this sum be thought of in any other way than as (—5) + (—5)?

[And so on.] Can you tell now what (—5) × 3 would be? (—5) × 8? (—5) × 11?

If 7 is subtracted from 0, what is the difference?

This represents the subtraction of how many 7's?

If from the difference obtained above another 7 is subtracted, what difference does that give?

How many 7's have now been subtracted?

Could this result be thought of in any other way than as —7 — (+7)? [Yes, as 7 × (—2).] Can you now tell what 7 multiplied by —5, 7 × (—5), would be? 7 × (—9)?

If —9 is subtracted from 0 what is the difference?

How many —9's have now been subtracted?

If from the difference above another —9 is subtracted, what is the difference?

How many 9's have we now subtracted?

How else might —18 be thought of? As 0 — (—9) — (—9)? As —9 multiplied by —2?

[And so on for a number of examples.]

Can you tell what the result will be of multiplying —9 by —7; i.e., (—9) × (—7)? (—9) × (—10)?

. . . .

What have you observed concerning the signs of the results?

Have you observed anything about the examples that would enable you to determine the sign of the result even before you multiply?

State your conclusions.

ALGEBRAIC SKILLS OF DIVISION

The typical expression for division in algebra is

$$\frac{a}{b} = c$$

where the number a is divided by the number b and the result is the number c.

From the fundamental associative skill of relation (see page 88) this expression means that a is equal to c multiplied by b, or

$$a = bc$$

Since bc is a product, the associative skills of multiplication apply to the signs of the numbers.

Four cases may be distinguished where signs of the numbers to be divided are involved:

$$(1)\ \frac{+a}{+b} = ?$$

$$(2)\ \frac{-a}{+b} = ?$$

$$(3)\ \frac{+a}{-b} = ?$$

$$(4)\ \frac{-a}{-b} = ?$$

In the first case the sign of the second number must be such that when the second number is multiplied by $+b$ the sign of the result is that of $+a$. The associative skills of multiplication (see page 54) show that the only sign that can produce a $+$ result when multiplied by a $+$ number is $+$. It therefore follows that the sign of the result of $\frac{+a}{+b}$ is $+$.

Likewise, the associative skills of multiplication indicate that the result of $\frac{-a}{+b}$ is $-$; that of $\frac{+a}{-b}$ is $-$; and that of $\frac{-a}{-b}$ is $+$.

Summary of Associative Skills of Directed Numbers in Division

The associative skills of multiplication apply directly to all cases in division so that when the numbers to be divided are both $+$, or both $-$, the sign of the result is $+$. When the signs of the numbers to be divided are one $+$ and the other $-$ the sign of the result is $-$.

CHAPTER VI

ASSOCIATIVE SKILLS OF INTERPRETATION

Algebra is a subject in which the pupil is constantly called upon to give new meanings to forms of language with which he has long been familiar, and to which meanings utterly foreign to algebra have already been attached.

Manipulation with algebraic symbols is often relied upon to give meaning to symbols and processes, but faith in exercise of this kind does not seem to be paralleled by results. If in the formula $a = lw$ a pupil can always solve correctly for l that ability seems to furnish little evidence that the operation has any more numerical significance than opening a window or adjusting a hat or making any other rearrangement of materials. Where $a = lw$, it may be possible to have $l = \frac{a}{w}$ as the result of a "shift" of position, and to have the solution mean absolutely nothing more than a *shift*. That this is true is indicated, if not proved, when after a year and a half spent in studying algebra, pupils will write the value of h in

$$A = \frac{1}{2}\,h(b + b')$$

in [1] twenty-six ways that are wrong. Out of 549 pupils 97, or 17.4 per cent, contributed to the incorrect results. Even those who solved correctly the equation given above demonstrated their weakness, so far as comprehension is concerned, by inability to formulate the equation for (not solve) the following problem:

> A ladder 20 ft. long reaches a window 16 ft. above the ground. Assuming that the ground is level, how far is the foot of the ladder from the front of the wall?

In this case 144 students, or 26.2 per cent, missed the problem, and of these same pupils, exactly 100 solved correctly for h in the equation given above.

[1] See Appendix, p. 107.

Further evidence that apparent solutions are sometimes of the nature of a shift of position, rather than the result of the application of principles is found in the fact that in the same group of 549 pupils only 6, or 1 per cent, were unable to solve correctly for l in the equation

$$A = \frac{Pl}{2}$$

but, as pointed out above, 97, or 16 times as many as failed to solve for l, were unable to obtain the correct value for h in the equation

$$A = \frac{1}{2}h(b + b')$$

although *the mathematical principles involved in these two equations are identically the same.*

If a pupil knew what he was doing, operations performed with symbols might give meaning to both symbols and processes, but when rearrangements alone are affected, the pupil neither *knows* nor *uses* any mathematical principles whatever. Although continued practice in rearrangement increases certain types of facility, there is little reason for belief that it increases understanding. To an element of appreciation that puts mathematical meaning into the use of algebraic symbols we shall attach the name *associative skill of interpretation.* For instance n as an algebraic symbol may stand for a great variety of meanings foreign to the subject and still take on every appearance of being used correctly in an algebraic expression. If n is a number, the question "What is twice the number?" may elicit the correct response, $2n$; but n may remain a letter, it may represent the word "number," it may stand for an identification mark of a something that fits the case, just as a certain monogram identifies a motor car, or it may stand for some object like "sheep." The expression "2 sheep" has some numerical implications, but its meaning is not pertinent to an algebraic number as usually employed.

Even to recognize n as a number is but a small beginning toward appreciating its full significance. In the expression $2n$, n is usually not only a number, but it is also a unit of value and is capable of representation as such in the number scale. (See page 37.)

If n is a unit of value, it should be noted that it may be either a primary unit, complete in itself, or an expression of the measure of a magnitude in terms of some other unit of measure. Thus, in the first sense, n may be a unit of length, a box, a quantity of electrical energy, or anything else to which measure or even the idea of existence may be applied. In this sense it is conceivable that n might represent merely the letter of the alphabet, though this would be an extreme case and, in most instances, it would be entirely misleading so to consider the symbol in any algebraic expression of quantity.

In the second sense which may be given to the interpretation of the meaning of n, its significance is that of a measured quantity of any kind, as, for instance, six dollars, half a mile, or one-fifth of a circumference. In all of these cases the primary unit of measure is something other than n itself, being, in the illustrations given, respectively the dollar, the mile, and the circumference. But whether n is a primary unit or not, its use as a unit of expressed value in an algebraic expression is of the same nature in every instance.

When used as a number, n takes on every characteristic that may be assumed by any number; it may be concrete or abstract, positive or negative, integral or fractional, and so on.

Looked at as a number, n may be an element of any group of numbers which enter into an algebraic expression, or n is a number element in any algebraic operation; thus in $n + b$, n is some number to be added to b. (It should be observed that n has been employed in the preceding discussion to represent any algebraic symbol whatever; b, therefore, takes on any of the attributes of n.)

The term $n + b$ itself is also a number and may be treated as such the same as though the value represented by $n + b$ had been expressed by a single symbol: $2(n + b)$ is such a use of $n + b$, where this number is multiplied by 2. That juggling with letters is often mistaken for the exercise of mathematical principles is indicated by the results given in Appendix of this study, Section II, page 106. An examination of the work done throughout the year and on final examination by more than six hundred pupils seems to furnish conclusive evidence that very frequently pupils study algebra for one and two years without actually realizing that they are dealing with numbers.

PRIMARY MEANINGS AS CONTRASTED WITH ALGEBRAIC PROCESSES

Each one of the fundamental operations has two or more aspects: that of an indicated operation and that of the operation performed. The sum of $2a$ and $3b$ is $2a + 3b$, the *operation* consisting in admitting that the sum is the sum; but the sum of $2a$ and $3a$ is not only $2a + 3a$; but also $5a$. In the latter case there is actually some operation to be performed. The cultivation of ability to distinguish between an indicated operation and an actual operation is probably an associative skill of the highest order of importance in algebra.

In subtraction the same kinds of situations are encountered as in addition.

In multiplication, the multiplication of a and b is completed by admitting that multiplication gives a product and writing or thinking the proper phraseology of a product, which is $a \times b$ or, as more frequently expressed, ab; but the product of $5a$ and $6b$ presents still another step; for $5a \times 6b$ equals $30ab$; while the product of $5a$ and $6a$ goes still further and yields $30a^2$.

The operation involved in changing $a \times a \times a$ to a^3 is one of counting. There is no way of proving that $a \times a \times a = a^3$. The two forms are merely different ways of expressing the measure of the same quantity.

On the other hand, once we have given to a^3 its meaning, we are able to express the value of the product of a^3 and a^4 and to *prove* that it equals a^7.

Likewise there is no operation involved in expressing x^{-2} as $\frac{1}{x^2}$ or $\sqrt{x}$ as $x^{\frac{1}{2}}$ except the operation of writing. Whatever skills there are in recognizing the two forms as equivalent are of the nature of definitions, not of reduction from one form to the other by an application of mathematical laws or processes.

THE GENERAL SCHEME OF NUMBER EXPRESSION

The way in which number values are expressed in algebra begins to differ sharply from the arithmetical method when more than one symbol is employed. The meaning of ab is a matter of multiplication; each factor affects the other in making up the value of the number. In arithmetic, the meaning of 26 is derived by a process of addition. Both digits are necessary to the number, but they operate alone. If we add 20 to 6 we get the number, but in ab if we add a to b we obtain something totally different from ab. The result of all this is that in such a number expression as $14xy$ we have two distinct ways in which the separate symbols contribute to the number. Fortunately the arithmetical or Arabic-numeral portion of algebraic numbers does not change either in form or in interpretation as the pupil passes from one subject to the other, but wherever either multiplication or division is applied to an algebraic number, the distinctions that have been pointed out must be observed.

If $14xy$ is to be multiplied by $4x$, the 4 and the x of the two numbers have different kinds of meanings whose significance cannot be appreciated without the associative skill which recognizes x as a *factor,* while 4 assumes an office not as a factor, in contributing to the value of the coefficient of $14xy$; but the coefficient as a whole enters into the expression as a factor.

Besides the number types illustrated by a, abc, and $4xy$ there remains to be considered those with exponents, appearing in the form of a^3, $7a^2$, and $6a^2b^3$. The rôle fulfilled by a number appearing as an exponent is that of indicating an operation. The 3 in a^3 means something totally different from the 3 in $3a$, and something still different from the 3 in $32a$. In the last two numbers the 3 is itself combined with a by a mathematical operation (that of multiplication) to make up the value of the number. In a^3, on the other hand, the 3 is not combined with the a in any manner whatever: a^3 has a value that depends entirely upon a itself; the 3 neither is added to, subtracted from, multiplied into, nor divided into a. The 3 plays somewhat the rôle of a governing force, merely taking account of and determining the way in which a acts.

Pupils who are confused with regard to the measuring of exponents as contrasted with coefficients undoubtedly contribute heavily in supplying the 28 types of errors found in simplifying

$$\frac{16a^3 - 8a^5}{4a^2}.$$

(See Appendix, p. 106.)

Reference has already been made to the fact that such a combination of symbols as $n + b$ may represent a number instead of two or more numbers. Wherever a sign of aggregation occurs such an interpretation is always placed upon the expression. This is a part of algebraic language. In the expression

$$a - (2x + b - a)$$

the quantity within the parenthesis is treated as a number and the expression should be interpreted as calling for the subtraction of $2x + b - a$. The bar separating the numerator and denominator of a fraction operates in precisely the same manner as does any other sign of aggregation. Failure to recognize this fact was apparent in the errors found in the reduction of the fraction discussed on the preceding page.

The same pupils who solved for l in $A = \frac{Pl}{2}$ with almost perfect accuracy (99 per cent) and only two types of errors, found 28 different (wrong) results for $\frac{16a^3 - 8a^5}{4a^2}$, and 10.9 per cent of the pupils made the wrong response. If we attempt to analyze the discrepancy, force is given to the claim that mere manipulation in the guise of a change of form, not involving mathematical operation, often leads to deceptive estimates of a pupil's algebraic abilities.

$A = \frac{Pl}{2}$ is a form sufficiently short to be remembered. It can be put into acceptable form without any algebra whatever; but $\frac{16a^3 - 8a^5}{4a^2}$ presents a real problem, into which meanings have to enter, because of its length. The results obtained seem to indicate, on the part of very many pupils, such weaknesses in dealing with the fractions as to throw considerable suspicion upon an assumption of real ability in the case of the formula stated, notwithstanding the apparent perfection or technique on the part of many pupils.

Fundamental Operations with Algebraic Numbers

In arithmetic the fundamental notion is one of operation with numbers. In algebra the fundamental idea is one of relation, followed by processes. Accordingly, the meaning of algebraic processes is very largely included in the forms of expression themselves. For the most part, to be able to *read* the language of algebra is to understand its operations.

ADDITION

Ability to add is preëminently one involving associative skills of interpretation, because there are three distinct meanings given to this operation in algebra, whereas there is only one in arithmetic.

The first type of algebraic addition is one into which the addition of coefficient enters. Such a problem is encountered in the addition of $-2ab^2$, $6ab^2$, and $5ab^2$.

The second type is one with which coefficients have nothing to do, and the operation can only be indicated. This type is illustrated in the addition of such numbers as x, 4, and pq.

The third type calls for the addition of numbers whose form must be changed before the addition can be effected. Numbers belonging to this type include fractions and radicals. It is possible to add in the sense of the first type only after a change in form of such expressions as

$$\frac{a+b}{a-b} \text{ and } \frac{-2ab}{a^2-b^2} \quad \text{or} \quad (8x^2)^{\frac{1}{2}} \text{ and } (72x^2)^{\frac{1}{2}}.$$

The outstanding associative skill of the operation of addition is that which recognizes and distinguishes the three types that have been mentioned. There is involved the threefold capacity for recognition of likeness, difference, and the potentiality of algebraic expressions. So far as first-year algebra is concerned, expressions which are different only with respect to their coefficients are considered alike, and capable of being added according to the directed number characteristics of their coefficients. This is a purely arbitrary attitude, but apparently is sufficient for very elementary algebra.

In the examples above: namely, $-2ab^2$, $6ab^2$, and $5ab^2$, the entity to be added is considered as ab^2, the coefficients -2, 6, and 5 being the quantities of the entity to be added. Consequently, as in all cases of addition the total is the sum of the separate measures *of the article,* which amounts to $9ab^2$.

The associative skill which recognizes ab^2 as a particular number, must recognize $8a^2b$, $3a^2b^2$, and $-7b^2$ as *unlike* numbers, and therefore incapable of being added, in those forms, except by implication. Their addition, therefore, takes the form of $8a^2b + 3a^2b^2 - 7b^2$, which involves an element of adjustment not found in the $9ab^2$ above, in that the latter, in itself, contains a notion of finality of form not so easily apparent in the result which *indicates* addition.

The addition of fractions introduces a question, and some skills, relating to the meaning of units of measure that are foreign to the ideas of addition of integers. The subject of fractions may well be considered here, since, after all, the study of the fraction is a new topic with respect to the form of the symbols which are encountered, but not with respect to the operations that are employed. The treatment of the fraction demands a reorganization of its symbolism, but it does not require any new manipulative skill.

The problem:

$$\text{Add } \frac{a}{b} \text{ and } \frac{a}{d}$$

might be considered as satisfied by writing $\frac{a}{b} + \frac{a}{d}$. It is difficult to see how anything would be gained in this case by an attempt to carry the operation further, although that can be done.

The problem:

$$\text{Add } \frac{a+b}{a-b} \text{ and } \frac{-2ab}{a^2-b^2}$$

is of a type usually considered as capable of simplification of form beyond that in which the exercise is here stated.

Any simplification of fractions which stand in such a relation as do these two demands of the pupil whose experience has been confined to integers either new manipulative skills or some new associative skills capable of bringing into play the skills already prepared to act in the addition of integers.

Just as ab may be considered as the unit b taken a times, or $2(x+z)$ may be looked upon as a unit $x+z$ taken 2 times, so any fraction may be looked upon as a unit, taken a certain number of times, where the unit of fractional value has been obtained by *dividing* some primary unit. Thus in the arithmetical fraction $\frac{7}{8}$, the unit value of the fraction is $\frac{1}{8}$, obtained by dividing a primary unit 1 by 8. In the algebraic fraction $\frac{n}{d}$ the unit of fractional value is $\frac{1}{d}$, obtained by dividing 1 by d; n sustains the same relation to the expression $\frac{n}{d}$ that a bears to the expression ab, where this form indicates that b is multiplied by a. In the same way $\frac{n}{d}$ may be looked upon as denoting that the fractional unit $\frac{1}{d}$ has been multiplied by n. It is necessary to put into each of these forms these associative skills of interpretation, if they are to be used in operations with numbers.

Another skill of interpretation that needs to be taken into account in the addition of fractions has to do with the possibility of changing the form of a fraction *without changing its value.* This is no new idea. Such transformations are constantly made in addition and subtraction. For instance, when the terms of $6x^2y - 5x^2y + 2x^2y$ are added and the sum is stated in the form $3x^2y$, there is no notion that the values of either expression have been changed—whatever value the original expression had is merely stated in a new form. The most striking contrast between the significance of a change in form in a problem involving addition of terms as they stand and a problem requiring a change of form in a fraction consists of the fact that the change of form that takes place in a sum is final, while the change that takes place in a fraction, in order that it may be added, is preliminary to the actual addition.

The laws of fractions, "Multiplying or dividing both terms by the same number does not alter the value of the fraction," does not necessarily carry either conviction or a sense of its own intrinsic value to the mind of the pupil. Pupils who could perform with a high degree of accuracy all of the manipulative operations involved in simplifying $\frac{1}{a-b} - \frac{3a}{a^2-b^2}$ were still unable to effect the simplification in 84.9 per cent of the cases. (Problem 2, Appendix, Section I.)

A foundation for the associative skill which enables one to comprehend the change in form of a fraction is found in the use of different units of length, weight, and the like. Suppose that a certain distance is measured in terms of *feet.* It is possible to consider the same distance as measured in inches or $1/12$ of a foot. Whatever the number of feet there are in the distance, the number of inches is 12 times the number of feet, but the distance does not change. In the transition from feet to inches the unit of measure has been *divided* by 12, but the number of times the measure is taken is 12 times as great as before. The same kind of changes—dividing the unit and multiplying the number of times the unit is taken—occurs in the case of the fraction. If a certain fractional unit is $\frac{1}{n}$; that is, if 1 is divided into n equal parts, and one of these parts is used as a unit, compare this fractional unit with that which results if the 1 is divided into $2n$ equal parts. The former fractional unit equals 2 of the smaller units; therefore it is consistent to write $\frac{1}{n} = \frac{2}{2n}$, because $\frac{2}{2n}$ means $\frac{1}{2n}$ times 2. The same conclusions follow if the number by which the original unit is divided is multiplied by a. The measure of the fractional unit now becomes $\frac{1}{an}$, but the original fractional unit now equals a of the derived units, so that it is consistent to write $\frac{1}{n} = \frac{a}{an}$. Since $\frac{1}{n} = \frac{a}{an}$, where emphasis is placed on $\frac{1}{n}$ as the original fractional unit, the subject of the sentence may be made $\frac{a}{an}$ and the equation written in the form $\frac{a}{an} = \frac{1}{n}$.

A further associative skill requisite to the addition of two fractions is an awareness of the possibility of a common fractional unit, and an appreciation of the relation that the given units have to the desired one.

The final associative skill resides in an appreciation of the significance of the act of combining (adding) the fractions, after expressing them in terms of a common fractional unit. The numerators become, in effect, the coefficients of the fractional unit.

Thus, in adding the fractions

$$\frac{a-b}{a-b} + \frac{-2ab}{a^2-b^2}$$

the various associative skills lead to the forms:

$$(1) \qquad \frac{(a+b)\,(a+b)}{a^2-b^2} + \frac{-2ab}{a^2-b^2}$$

(2) $$\frac{a^2 + 2ab + b^2}{a^2 - b^2} + \frac{-2ab}{a^2 - b^2}$$

(3) $$\frac{a^2 + b^2}{a^2 - b^2}$$

In the second step $a^2 + 2ab + b^2$ is considered as a coefficient of the fractional unit $\frac{1}{a^2 - b^2}$, and likewise $-2ab$ is considered as a coefficient of the same fractional unit.

The addition of radicals involves no very extended treatment of the topic in first-year algebra, but there is required an understanding of the elementary theory of exponents. As has been pointed out in the chapters on Associative Skills of Appreciation, the meaning of x^3 is $x \times x \times x$, and for every expression of this kind the exponent is a shorthand way of indicating the number of times the expression to which it is applied is used as a factor. $a^m \times a^n$ may be written a^{m+n} because in the product $a^m \times a^n$, a is used as a factor $m + n$ times.

$\frac{a^m}{a^n}$ may be written a^{m-n} because the division of a^m by a^n is equivalent to taking out the factor a, n times; just as the division of 24 by 8 is equivalent to taking out the factor 2, 3 times.

$(a^m)^n$ means that a^m is to be used as a factor n times, so that $(a^m)^n = a^m . a^m \ldots .$ until a^m has been used n times as a factor, but this means that a will then have been used as a factor mn times. Accordingly $(a^m)^n = a^{mn}$; $(ab)^n$ may be written as $a^n b^n$, because $(ab)^n$ means that ab is to be taken as a factor n times, therefore $(ab)^n = ab \times ab \times ab \ldots$ until ab has appeared as a factor n times, but in that case a will have been used as a factor n times and b will have been used as a factor n times in the same product, therefore the conclusion:

$$(ab)^n = a^n b^n.$$

Since $a.a.a = a^3$ or $a.a.a \ldots$ to n factors $= a^n$, if we wish to think of the factor whose repetition 3 times has produced a^3 the form $(a^3)^{\frac{1}{3}}$ is used. This implies the question, *what factor has been used 3 times to produce a^3?* Likewise $(a^n)^{1/n}$ asks what factor has been used n times to produce a^n; $(8)^{\frac{1}{3}}$ asks what factor has been used 3 times to produce 8; $(16)^{\frac{1}{2}}$ asks what factor has been used twice to produce 16. A radical, then, states a product and requires, by implication, the determination of the factors of the product. It is possible to ask questions of this kind that cannot be answered in terms of whole numbers. For instance, $(2)^{\frac{1}{3}}$ asks what number used as a factor 3 times will produce 2. There is such a number, but it is an approximation and its determination is a subject for a more advanced course. We must here be content with merely expressing the idea, not the actual number.

Products which contain a factor repeated several times may contain other factors as well. For instance, the product of $2 \cdot 2 \cdot 2 \cdot 3$ is 24. The expression $(40)^{\frac{1}{3}}$ asks what factor has been employed 3 times to produce 40. Obviously the question cannot be answered completely as in the case of $(8)^{\frac{1}{3}}$, but it can be answered partially because the equation $2^3 \cdot 5 = 40$ shows that 2 has been used as a factor 3 times, although 5 has not. If we wished to answer the question implied in $(40)^{\frac{1}{3}}$ as far as we could we might conveniently consider $(40)^{\frac{1}{3}}$ as written in the form $(2^3 \cdot 5)^{\frac{1}{3}}$. From this we can see that to produce 40 the factor 2 has been used 3 times and that this product has been finally *multiplied* by 5. The complete number, then, which is used as a factor 3 times to produce 40 appears to be 2 multiplied by some other number, which when used as a factor 3 times will produce 5. That latter number we have seen is $(5)^{\frac{1}{3}}$. It therefore appears that $(2^3\, 5)^{\frac{1}{3}} = 2\ (5)^{\frac{1}{3}}$.

Whenever a number like $(40)^{\frac{1}{3}}$ has been changed in form to $2\ (5)^{\frac{1}{3}}$, or a number like $(18)^{\frac{1}{2}}$ has been changed to $3\ (2)^{\frac{1}{2}}$ it is said to be simplified, and if the radical factors are alike as in the problem:

$$(8)^{\frac{1}{2}} + (18)^{\frac{1}{2}} + (50)^{\frac{1}{2}}$$

which may be expressed as

$$(4 \cdot 2)^{\frac{1}{2}} + (9 \cdot 2)^{\frac{1}{2}} + (25 \cdot 2)^{\frac{1}{2}}$$

and in turn as

$$2\ (2)^{\frac{1}{2}} + 3\ (2)^{\frac{1}{2}} + 5\ (2)^{\frac{1}{2}}$$

the expression may be still further simplified by considering 2, 3, and 5 as coefficients and adding as we should any other expressed sum of like terms, obtaining $10\ (2)^{\frac{1}{2}}$.

The entire theory of radicals up to this time has been made to depend upon the notion of a product. The most difficult point to grasp is the rather abstract notion that a number of the form $a^{\frac{1}{3}}$ is one of the three factors which multiplied together will produce a, but the analysis is parallel to that of taking the product of, for instance, x^4 three times,

$$x^4 \times x^4 \times x^4 = x^{4+4+4}$$

which equals x^{12}.

In the same way

$$a^{\frac{1}{3}} \times a^{\frac{1}{3}} \times a^{\frac{1}{3}} = a^{\frac{1}{3}+\frac{1}{3}+\frac{1}{3}}$$

which equals a.

SUBTRACTION

The associative skills of subtraction of algebraic numbers are precisely those of addition, subject only to the meaning of subtraction as contrasted with addition.

MULTIPLICATION

Multiplication, in algebra, is a far more simple operation than addition or subtraction, because the result of the operation of multiplication requires only a number which contains all of the factors of the numbers that are multiplied. Thus the multiplication of

$$6\,a^2\,b \text{ by } 5\,a\,x\,y^2$$

means that the result is to be found in an algebraic expression that denotes multiplication (see Associative Skills of Interpretation, page 62). Such an expression would result were we to write

$$6 \cdot a^2 \cdot b \cdot 5 \cdot a \cdot x \cdot y^2$$

but this is not convenient; since the product of 6 and 5 is 30, and a is used as a factor 3 times it is better, and more usual, to express the product as

$$30\,a^3bxy^2$$

The multiplication of a polynomial by a monomial yields the result that is obtained by considering the polynomial as composed of monomials and applying the appropriate signs, as in the case of $2\,a^2 + ab - b^2$ multiplied by $3\,ax$. The multiplication of a polynomial by a polynomial may be considered as the algebraic sums of the results obtained by multiplying one polynomial by each of the monomials of the other. Thus the product of $a + b$ by $x - y$ is that which is obtained by multiplying a and b separately by x, and also multiplying a and b separately by $-y$ and combining the results obtained, according to the laws of algebraic addition.

Multiplication in which fractions are included in either of the numbers multiplied requires an associative skill of interpretation of the meaning of a fraction. The expression $\frac{a}{b}$ may be considered as meaning a divided by b. This is not the same interpretation of a fraction that was suggested in addition, but it is a better one for multiplication, because addition of fractions generally requires a change in their form. Multiplication makes no such demand on the form of the fractions to be multiplied.

Therefore multiplication of $\frac{a}{b}$ by x yields $\frac{ax}{b}$, because this expression retains a and x as factors and retains b as a divisor. (See also Addition, page 65.)

The multiplication of $\frac{a}{b}$ by $\frac{x}{y}$ yields $\frac{ax}{by}$, because this expression retains a and x as factors and b and y as divisors. (See also Addition, page 66, concerning the multiplication and division of a fraction.)

DIVISION

Division is the reverse of the process of multiplication in every particular. In multiplication two factors are given and the problem is to find the result of multiplication. In division the result of the multiplication of two factors and one of the factors are given, and the problem is to find the other. These are basic facts which occasionally may be considered as throwing light on the operation of division.

By far the most frequent and most important use of division in algebra is in connection with factoring. Factoring demands of the pupil the ability to see in a result of multiplication; that is, to see in the number to be factored two or more factors. In a monomial like $6a^2b^3x$ there are many factors; for instance 6, a^2, b^3, and x are factors; or 2, 3, a^2, b^3, and x are factors; or 6, a, a, b, b^2, and x are factors. In fact, *any* array of factors which multiplied together will result in $6a^2b^3x$ constitutes the factors of the number. The factoring of a monomial factor from a binomial or a trinomial calls for the ability, first, to factor each term of the polynomial and, second, to select the factors which are alike. The same kind of a skill is demanded here which is required in determining those qualities which permit two terms to be added (see Skills of Addition, page 65).

The manipulative skills of factoring binomials and trinomials may be placed on an arbitrary basis relating to types such as $x^2 - y^2$ and $ax^2 + bx + c$. The associative skills, however, are based upon experience in multiplication. This experience necessitates taking note of the relation of the numbers multiplied to the product, yielding in the case of multiplication of:

(1) a polynomial by a monomial, a polynomial of the same number of terms.
(2) a binomial by a binomial, either
  (*a*) a binomial or
  (*b*) a trinomial.

This experience with multiplication leads one, then, to expect either a monomial factor, or in the case of a binomial or a trinomial no factor other than a monomial or binomial.

Division, not of the nature of factoring, that is, in the case where both numbers to be divided are expressed, falls into two types:

1. Division indicated as in a practical expression of the form $\frac{x}{y}$ in which x and y may or may not be of the form $\frac{an}{ad}$ (see discussion of change of form of a fraction under *addition*, page 68).

2. Division actually performed. This latter type of division is dependent upon considering the number to be divided as obtained by multiplying each term of the division by some other number. (Even though the number to be divided has not actually been obtained in this manner: that is, when there is a remainder, the process of dividing proceeds upon this assumption.) Division, then, is an orderly process of determining by what number the divisor might have been multiplied to produce the number that is divided.

Division is a process that may in all cases be dominated by the idea of multiplication.

Summary of Associated Skills of the Fundamental Operations

ADDITION

1. Addition demands recognition of *likeness, unlikeness,* and the *potentiality* for likeness of algebraic numbers.

2. The addition of fractions is dependent upon the possibility of expressing two or more fractions in terms of the same fractional unit. To which are subsidiary:

(a) $\frac{n}{d} = \frac{a\,n}{a\,d}$

(b) $\frac{a\,n}{a\,d} = \frac{n}{d}$

where, in each case the fraction is considered as being stated according to the first member.

3. An awareness of the particular form which the fractional unit may take.

4. The recognition of the meaning of fractions expressed in terms of the same fractional unit.

5. In the addition of radicals, the meaning of $(a^p\, b^q)^{1/n}$.

6. The significance of the laws of exponents:

(a) $a^m \times a^n = a^{m+n}$

(b) $\frac{a^m}{a^n} = a^{m-n}$

(c) $(a^m)^n = a^{mn}$

(d) $(ab)^n = a^n b^n$

(e) $\left(\frac{a}{b}\right)^n = \frac{a^n}{b^n}$

7. The recognition of what constitutes likeness and unlikeness of radical expressions.

SUBTRACTION

The associative skills of subtraction of algebraic numbers are identical with those of addition, except for the definition of *subtraction,* as distinguished from the definition of *addition.*

MULTIPLICATION

1. The meaning of *multiplication* in algebra is contained in the idea and form in which algebraic numbers are expressed.

2. The multiplication of polynomials may be considered in terms of monomial multiplication.

3. Multiplication of fractions is interpreted in terms of the meaning of fractions, as either a fractional unit, or as an indicated operation of division.

DIVISION

1. Factoring is a form of division, and is accomplished by considering the expression to be factored as a result of the multiplication of two or more numbers.

2. Besides factoring, division includes two types of operations:
   (*a*) The fractional, indicated form.
   (*b*) Actual division, when the operation is carried farther than in (*a*).

3. Actual division of two numbers is directed by the idea of multiplication of the divisor by some number to produce the number that is to be divided.

The Axioms

The axioms are no longer considered as *a priori* facts. To the mathematician they are convenient assumptions upon which to base mathematical reasoning. To the psychologist they are succinct generalizations of experiences with quantities. To the pupil whose acquaintance with quantitative measurement has been confined chiefly to operations with measures instead of relations between measures there have come comparatively few opportunities for reaching axiomatic conclusions. The notion of equality and of inequality is present whenever a choice is made between two alternatives, but such ideas are in terms of estimates which are generally influenced by *quality* as well as *quantity*. The mathematical axioms operate only in the presence of concepts which are expressed in the abstract language of exact measurement. Time and experience with many events in which the behavior of equal quantities is observed in a scientific manner are necessary in order to acquire a feeling for the significance of statements, the derivation of whose meaning is wholly dependent upon the way in which quantities *have* acted when compared one with another.

These are the reasons for not considering the axioms as contributing initially to the associative skills which enable a pupil to answer the questions or refine the facts of an equation. Growth finally eventuates not alone in the ability to deal with more difficult situations, but also to discard some of the details of earlier modes of thought and procedure and to replace these details with conclusions which are the outgrowth of the organization of earlier associations.

The associative skills of relation which have been developed for the equation provide for every contingency that might arise in connection with the equation of the first degree. For equations of higher degree a like set of associative skills might be found, but it would be contrary to good sense to continue indefinitely to operate only with skills which are fundamental to the beginner. The associative skills that have been developed will always apply economically to the equations with which they have been illustrated. Their exercise has, however, afforded the background for the development of further skills whose operation takes the form of axiomatic statement and procedure. Comprehension of the meaning of the axioms is thus made to grow out of, and be a part of, the ability to generalize, which it is the peculiar function of algebra to develop.

The Function Concept

No discussion of the associative skills of algebra would be complete without reference to the *function concept,* but it is a topic that is difficult to classify. It is not a topic of algebra that may be segregated from others, like *addition of monomials,* or *solution of linear equations in one variable,* or *the multiplication of directed numbers.* The *function concept* is among the most important of all algebraic ideas, and it extends through every part of the subject, wherever there is a number expression capable of assuming many values. It might be discussed under skills of relation, but that chapter is reserved for relations implied by the sign of equality. The idea of functionality is not confined to the stated equation, although it is inherent in many equations.

Whenever the value of a numerical expression varies according to a law, the numerical expression is said to be a function of the law under which it changes. A simple illustrative example is found in the equation $y = 2x$. In this expression the value of y varies according to the values assumed by x, or as we say, "*y is a function of x.*" But the expression $2x^2 - 3x + 4$ is also a function of x; although no equation is here expressed, the expression varies according to changes in the value of x, and therefore fulfills the definition of a function.

Every formula contains the idea of a function, because there are always at least two variables in a formula, and each is affected by the values of the other.

The associative skill of the function concept is an awareness (1) of the presence of a number expression containing a variable; and (2) of the fact that the value of the expression itself is dependent upon, and changes simultaneously with, changes of value in the variable.

Summary of Associative Skills of General Interpretation

1. The recognition of the number implications is an important part of the understanding of algebraic processes and relations.
2. Algebraic forms of number expression differ from those of arithmetic and assume definite meanings according to the way in which they are written; as ab, $\frac{a}{b}$, $a + b$, and so on.
3. The function concept extends to all number expressions in which a variable appears, and consists of an awareness of value in the expression which depends upon the variable.

The Line Graph

The line graph is an ingenious appropriation of the idea that on a plane a point may be thought of as being moved from one position to any other by moving in but two directions, by two definite amounts. This is a familiar fact. We commonly get from place to place by moving east and south or east and north, and so on. If there is nothing in the way two directions and two distances are always sufficient. The graph simply lets each of these distances stand for the measure of some quality that is to be expressed. The beginning of a graph, then, is the establishment of some fixed places from which each distance and direction may be measured. This is usually accomplished by taking two perpendicular lines.

It is not at all necessary that a graph shall express *two* conditions (one may be denoted by separate lines of appropriate length), still the two-condition graph is the one that invites special consideration from an algebraic standpoint.

There are two fundamentally distinct conceptions of the graph: (1) the mathematical graph; and (2) the statistical graph. These terms are not entirely satisfactory, because a statistical graph may be, and often is, mathematical; we will use these words here in a technical sense, with meanings as indicated in the following pages.

A mathematical graph is one in which a fixed mathematical relation exists between two variables, but only two. For instance, it is possible to construct the graph of such relations as are expressed by $x - y = 2$, $y = x^3$, $y - x^2 = 4$, and so on. It is not possible in a plane, to construct a graph of $x + y + z = 1$ or of $xyz = 2$, and so on, because the three variables here denote that the position of a point must be fixed by three conditions, an impossibility in the plane.

As explained on page 89, such an expression as $y - x = 2$ is satisfied whenever y and x take values such that $y - x$ equals 2. There is an infinite number of such pairs of values, which suggests that the graph must indicate that fact. The equation in this case is satisfied when y and x take the pairs of values indicated by the horizontal lines in the accompanying scheme.

	x	y
A	0	2
B	-2	0
C	1	3
D	2	4
E	-1	1
F	5	7

The arrangement by which the graph, or picture of the relations between these pairs of values is indicated is entirely arbitrary but convention has become practically standardized in this country, so that the accompanying diagram is representative of general practice. Geography has probably already laid the foundation for the idea that the position of a point is fixed if its distance east or west of a certain line and its distance north or south of another line are given. This is the plan that is followed in the graph, except that the perpendicular lines are called axes and the distances expressed in terms of $+x$ and $-x$, and $+y$ and $-y$. (See Chapter V, on Directed Numbers for the direction meaning of $+$ and $-$.)

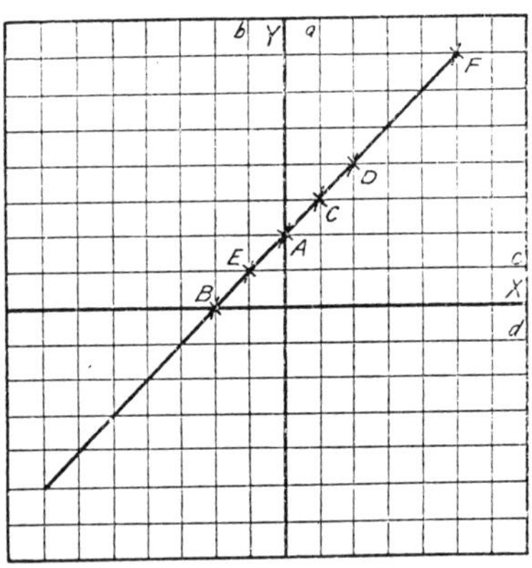

Any point on line a is considered as of a distance $+ 1$ (unit of measure) from the Y-axis in a $+$ direction; any point on the line b is located at a distance of $- 1$ from the X-axis, and so on. A point on line c is considered as located at a distance of $- 1$

from the X-axis and a point on line d at the distance of —1 from the X-axis, and so on. The *location* of a point is thus made to depend upon the elementary ideas of distance and direction, and, for graphic purposes, the position of the point is translated into whatever convenient meanings may have been attached to *distance* and *direction.*

In this case the position of D, reading from the graph, indicates that when x equals 2, y equals 4; the position of E that when x *equals* -1, y equals $+1$, and so on. The interpretation of the graph puts back the meanings that were initially introduced into its construction.

From the equation, by means of substitution of either of the two values of the horizontal lines of the diagram, the other value is obtained. Thus, when $x = 0$, $y = 2$, and so on. Any number of points, besides those indicated might be found. The straight line which passes through the points is merely an indication of where we think the others would be if we could locate all of them.

That the associative skills of graphic interpretation may be readily learned is indicated by the result of the examination of 549 pupils, to which reference has previously been made. (See Preface.)

Eight problems involving both construction and interpretation of graphs were assigned. The results were so uniform as to indicate that the pupils' interpretation of meanings was not obscured by a manipulative process of construction.

The eight problems which were presented to the pupils were as follows:

1. In order to find out whether the system

$$x + y = 11$$
$$x = y + 5$$

is solvable I have graphed this pair of equations on this section of squared paper and I find that . . . .

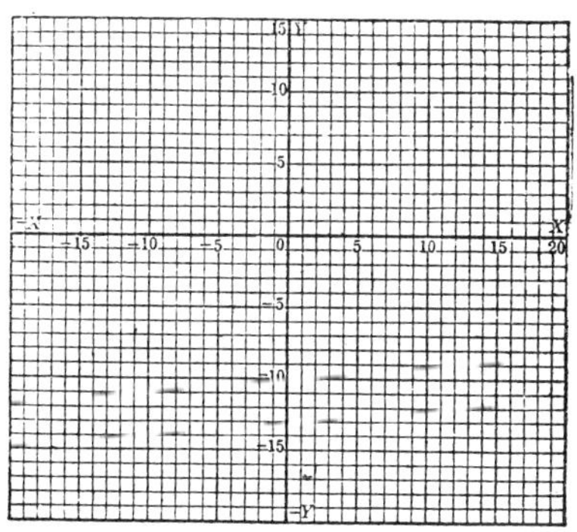

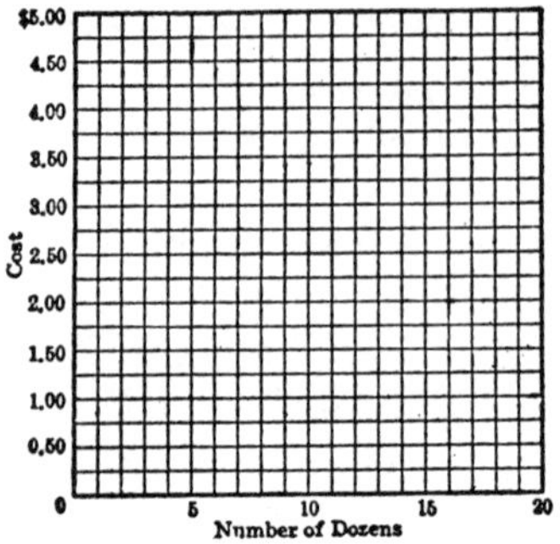

2. On the accompanying section of squared paper draw a cost graph for peaches at 20 cents per dozen.

3. The graph in Example 2 shows that the cost of 7½ dozen peaches is $........

4. The graph in Example 2 shows that .... dozen peaches can be bought for $2.80.

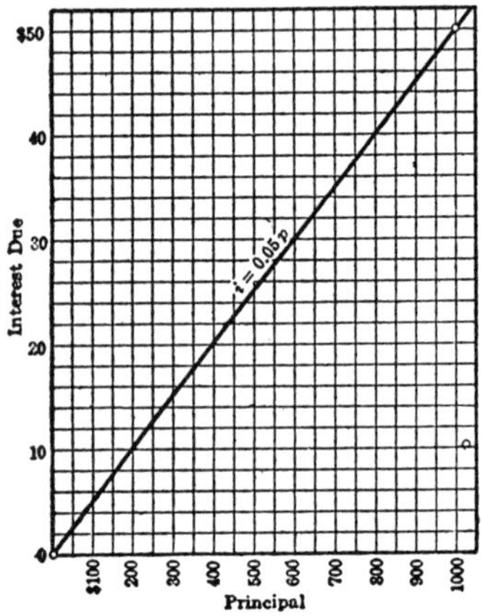

5. The graph of the formula $i = prt$ may be used in finding the amount of simple interest due on investments at different rates of interest. For example, the accompanying graph of the formula $i = 0.05p$ shows that the interest for 1 year on $300 is $........

6. The graph in Example 5 shows that the interest at 5 per cent for 1 year on $300 is about $........

7. The graph in Example 5 shows that if I wish to receive $12 a year in interest I must invest about $........ at 5 per cent.

8. The graph in Example 5 shows that if I wish to receive $36 a year in interest I must invest about $.......; that is, the interest received depends upon the ....

It will be observed that these graphs were skillfully designed to test a good many things, at least the following abilities being demanded:

1. The ordinary graphic approach to the solution of systems of equations, including:

(*a*) Construction of the graph.
(*b*) Interpretation of the result. } (Example 1)

2. The construction of a graph from informational data. (Example 2.)

3. The interpretation of the meaning of a graph constructed by oneself in terms of the general and particular information that it conveys. (Example 3, Example 4.)

4. Ability to read a graph that is presented. (Example 5, Example 6, Example 7.)

5. Ability to generalize as to the implications of a graphic representation. (Example 8.)

Among pupils who possessed the associative skill of interpretation in terms of the significance of twofold conditions covering the location of a point, we should expect to find a fair degree of consistency in the facility with which these different questions were treated. Had the ability been upon a purely manipulative basis with the separate questions representing separate jobs there would have been introduced the possibility for great variation in the results obtained for the different problems.

The accompanying table shows a remarkable uniformity in the results. To be sure, this might have been the effect of intensive manipulative training, but such explanation is rendered unlikely by the inclusion of several schools and the fact that no advance information was obtainable by the schools regarding the nature of the examination for which to prepare.

RESULTS OBTAINED BY 549 PUPILS FROM 35 SCHOOLS IN CONSTRUCTING AND INTERPRETING GRAPHS

	PROBLEM NO.							
	1	2	3	4	5	6	7	8
No. of times solved correctly	489	484	480	449	451	493	470	478
No. of times solved incorrectly	52	42	43	78	59	18	38	31
No. of times omitted	8	23	26	22	39	38	41	40
Per cent of correct solutions	89.0	88.1	87.4	81.7	82.1	89.8	85.6	87.0

Although the per cent of correct solutions for problems 1, 2, and 3 is very nearly the same, an indication that the ability to read a graph is not necessarily implied in the ability to construct one is to be found in the fact that 30 of the pupils who failed to interpret No. 3 correctly had constructed No. 2 properly.

The sudden drop in the per cent of correct solutions when No. 4 is reached may be due to the common practice of regarding y as a function of x in ordinary classroom procedure and in following this convention by more frequently reading and considering the values of y corresponding to certain values of x, than in considering the values of x as also determined by those of y. The confusion that attended No. 5 is a fair indication that attention was well centered on the graph instead of the simple arithmetic of the correct response.

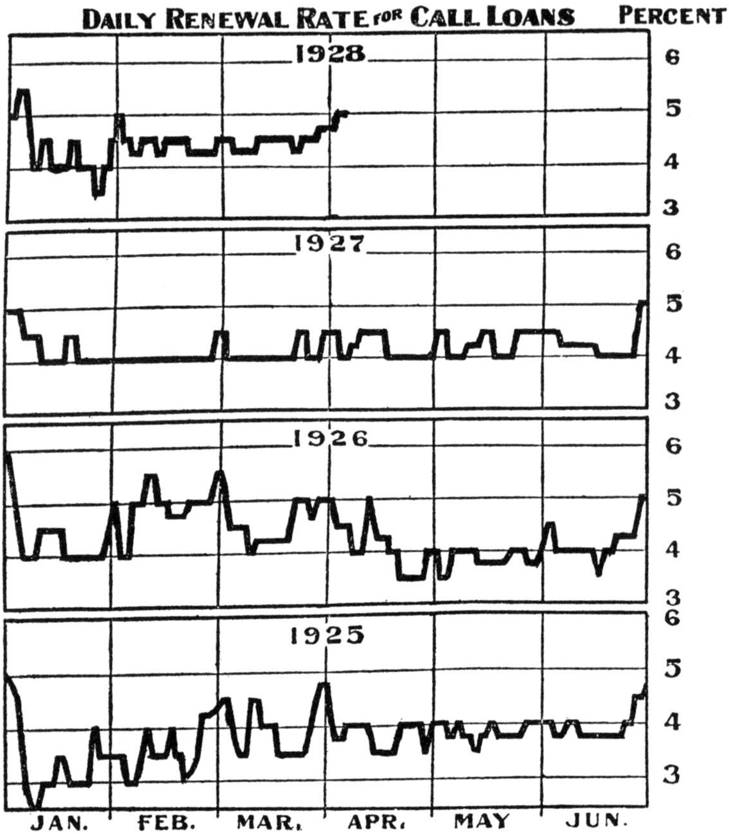

Such graphs as have been described are functional mathematical graphs in the strictest sense of the word. The position of each point on the graph is fixed by the operation of the same law: In the case of the peaches that law is that the price is 20 cents a dozen, which means that the price of a supply of peaches depends upon the number of dozen purchased. In the interest problem, the interest is five cents on the dollar, which means that the total interest depends upon the number of dollars at interest. Each of these cases illustrates the meaning of *function,* as discussed on page 76.

Another and very different type of graph, from a mathematical standpoint, is the one that has been called a statistical graph. This kind of graph is in all respects the same as the functional mathematical graph, except that there is no idea of the dependence of one quantity upon another. Such graphs are illustrated by the accompanying diagram, which indicates changes in money values from time to time. Here a mere fact is revealed, which happens to be true. That is, the interest on loans at any time is not dependent in any way upon that particular date, but is due to other circumstances. It will be observed that these graphs in each case are not constructed on a scale that represents the real comparative distances of their points from a reference line. They are so-called amputated graphs, necessitating an evaluation in terms of elements that are not present. Pseudographs would be a good name for such representations, and their evaluation demands a skill of interpretation of implication as well as direct reading.

Summary of Associative Skills of Line Graphs

1. Construction and reading of line graphs is the exercise of the fundamental notion of motion in a plane, which is that all points may be reached from a given point by moving in not more than two directions.
2. Interpretation of pseudographs is in terms of statements both directly and indirectly implied.

CHAPTER VII

THE EQUATION AND RELATED SKILLS

The oldest, and probably the most important, expression of relation in algebra is the equation. Its definition as "a statement of equality between two quantities" necessitates a rather strict interpretation of the meaning of the word *quantity,* and also requires one to distinguish between the use of the word equation in algebra and the proper but less specific use of the word in arithmetic. For instance, in arithmetic if we wish to express the cost of six tickets at $1.65 each we may say, or write

(1) The cost of the tickets $= 6 \times \$1.65$.

If we contrast this with an algebraic equation, say,

(2) $x + 2 = 9$,

we immediately encounter some striking dissimilarities. In the first equation the sign of equality has a significance quite different from that of the second equation. In the first instance the symbol $=$ is merely a convenient way of connecting "the cost of the tickets" with the fact that this cost is obtained by multiplying $1.65 by 6; the symbol has no distinctive mathematical significance as to either implication of functional relationship or any mathematical operation; a colon would do as well.

In the second equation the symbol $=$ means exactly what it says: namely, that there is a condition of equality between the two *numbers* $x + 2$ and 9.

The most striking dissimilarity between the two equations is not connected with the symbol $=$ but concerns the interpretations that have to be made in solving the equations. In (1) there is really no interpretation called for; the equation is solved when the indicated operation of multiplication, and nothing more, has been performed. This is far from being the case in (2). Here the only apparent operation is that indicated by the sign $+$ but this is the sign of precisely the operation that is *not* performed practically in order to solve the equation.

Two methods of solution of the equation that have been most

extensively employed have failed to produce adequately satisfactory results. The "transposition" method has been discredited because it neglects any semblance of mathematical procedure. Application of the axioms is, to the ordinary pupil, a use of formal rules. If either of these methods succeeded we should not be likely to find the confusion that exists in the treatment of the four equations discussed on page 16, and other instances with which every teacher is familiar.

There remains a third method of attack that subordinates manipulation—the performance of any operation—to an understanding of the meaning of the relations expressed between the numbers appearing in the equation. This method is one which would employ the manipulation of algebraic operations because they are naturally associated in the pupil's mind with the results he desires, while by the first two methods the operations are performed with more or less blind disregard of their significance, but with faith that their consequences will be reliable. The extent to which manipulation, to the exclusion of meanings, dominates ordinary procedure with equations was demonstrated by writing the portion of the equation

$$x + 2 =$$

on one side of a sheet of paper and submitting it to 76 adults with the question, "What do you know about the second member of the equation, if solving for x?" The answers varied from, "Don't know anything," "Might be anything," "Doesn't contain x," to "You will have to see the second number before you can tell." So far as could be determined only 5 of the 76 persons interrogated had even thought of the real relation between the members and x that is absolutely expressed by the first member alone. Every one of the persons who took part in the test agreed when the fact was pointed out to them that

$$x + 2 =$$

as it stands by itself means *the second member of the equation is 2 more than x.*

The recognition of this fact is what we call an *associative skill.* The application of this skill to this particular equation at once reduced the solution of the equation to the elementary concepts of "more than" or "less than." The second member of the equation, whatever it is, becomes "2 more than x." The determination of x then becomes the kind of process with which the pupil has met in arithmetic and everywhere else. Transposition and axioms may later enter to abbreviate the actual work, but not to complicate the pupil's thinking; he has already done the fundamental thinking when he has perceived the essential fact that the second member of the equation is more than x by the amount 2.

To put a pupil through the task of getting x from a large number of equations, with the hope that he will stumble upon the *understanding* of the process seems quite as likely to confirm him in the exercise of an uncomprehending routine as to lead him to *generalize* or to approximate any of the algebraic virtues.

Two distinct types of significance attach to every equation, one type belonging to the purpose for which the equation is to be used, the other belonging to the relations between the symbols employed. In the first sense the meaning of the equation changes with each new setting in which it is encountered; the equation

$$y = \frac{9x}{5} + 32$$

when encountered in the algebra class may denote something totally different from the equation

$$F = \frac{9C}{5} + 32$$

as encountered in the physics class. So far as primary meanings of the symbols are concerned, that seems necessary and inevitable; but when the appreciation of the numerical relations involved and the ability to solve for x in the first equation carry no sure intimation of ability to solve for C in the second, then something has been left out of the pupil's comprehension of the meaning of the equation, or else the idea of algebraic generalization is wrong.

Drill, repetition, dealing with the same equation with changed symbolism are necessary to give facility and a sense of ease in handling such situations, but educative devices demand that the pupil shall work to some purpose [1] if he is to gain the greatest

[1] Thorndike, E. L., *Educational Psychology*, Vol. II, p. 51. Bureau of Publications, Teachers College, Columbia University, 1925.

return for the effort expended. Of course this general psychological law is operative in algebra, if it is anywhere, but what is the purpose? Certainly not merely to get the result. Quite as certainly exercise is for the purpose of inculcating awareness of a process, but why should we leave the attainment of this awareness to chance? Why not inform the pupil of the purpose at once and then let him act with a view to illustrating the main purpose?

In solving the equation $3x = 15$ there is a reason for "dividing both members by 3," but the reason is not because "equals divided by equals gives equals." The real substance of the statement resides in the fact that the expression $3x =$ means that the second member of the equation is 3 times x. The recognition of this fact is the *associative skill* of this relationship. The exercise of this skill brings the operation of solution down to the applicaton of elementary notions; namely, that 15 is 3 times some number, x, and that if a given number is 3 times a required number, that required number is obtained by dividing the given number by 3. This is difficult enough—very likely even more difficult than following the injunction "divide both members by 3," but the associative skill employs the idea of the meaning of the equation; the "divide-both-members-by-3" method requires nothing but an act—there may be no meaning whatever atached to the act.

Since associative skills cannot be illustrated by any manipulative device and are not themselves capable of symbolic representation, but rather the skills of association represent a property or characteristic that is defined by symbols and is antecedent to any operation, there is no shorthand way of expressing the skill that is analogous to that of expressing the equation. For instance, "solve $3x = 15$" is a job. "Solve $7z = 28$" is a different job, but *there is no difference in the associative* algebraic skills. Whatever differences there are are wholly of an arithmetical nature. In arithmetic comprehension of the facts involved in recognizing 5×3 as 15 and 4×7 as 28 are both separate jobs and separate skills. In algebra generalization plays no part in determining, and has no interest in limiting, the number of jobs; if the number of skills is to parallel exactly the number of jobs, generalization ceases to be an aim or even a possibility. The very existence of algebra seems to depend upon limiting the number of associative skills to a class numerically less than the number of jobs.

A principle, either of mathematical law or of organization of teaching material, which replaces attention to the details of each separate situation with generalized conclusions that are valid in individual cases, affords great economy of energy and practice.

The expression $x^2 + x + 41$ appears to yield prime numbers for every integer that is substituted for the variable but in reality this is a device in whose operation there is no certainty either of law or generality, and its validity is discredited by the fact that when 41 is substituted for x the resulting number is not prime.

On the other hand the fact that $a^3 \times a^2 = a^5$ is an illustration of a general truth which establishes the law $a^m \times a^n = a^{m+n}$ as well as would a million cases.

Associative Skills of Types of Linear Equations in One Variable

In the following discussion equations of the form $x + 2 = 9$, $2 + x = 9$, $9 = x + 2$, and $9 = 2 + x$ will be treated as representing a single type, and this plan will be followed generally. That is because we are discussing the associative skills which belong exclusively to the *solution* of the equation. To solve an equation there must also be exercised every skill of Chapters V and VI.

Type of Equation	*Associative Skill*
1. $x + 2 = 9$	The second member of the equation is 2 more than x. In this instance, 9 is 2 more than x.
2. $x - 2 = 1$	1 is 2 less than x.
3. $3x = 7$	7 is 3 times x.
4. $5x = 2x + 4$	Here there are at least three relations suggested by the conditions of the equation: (1) The second member of the equation is 5 times x; (2) the first member of the equation is 4 more than $2x$; (3) the first member of the equation is $2x$ more than 4. The exercise of (1), (2), and (3), yield respectively: (*a*) $x = \frac{2x + 4}{5}$ (*b*) $5x - 4 = 2x$ (*c*) $5x - 2x = 4$, or $3x = 4$. The third leads to a more simple equation than the given one, and hence is the skill to develop.
5. $\frac{x}{3} = 5$	The second member of the equation is ⅓ of x.

Type of Equation	*Associative Skill*
6. $\frac{3}{4}x = 11$	¾ of x is ¼ of $3x$. The second member of the equation is ¼ of $3x$.
7. $3x - 4 = 7x + 9$	A choice of associative skills is presented as follows: (1) The second member of the equation is 4 less than $3x$; (2) the second member is $3x$ more than -4; (3) the first member is 9 more than $7x$; (4) the first member is $7x$ more than 9. Which skill to employ is to be decided by the effect which its exercise has on the solution of the equation, and what this effect is can only be determined by trial. Judgment is the result of experience, and judgment is secured by actually contrasting one method of procedure with another. In this case the skill (1), whose exercise produces $3x = 7x + 9 + 4$ or $3x = 7x + 13$, followed by the exercise of the skill of the type (4), whose exercise produces $3x - 7x = 13$ or $-4x = 13$, yields a simplified form of the equation that is capable of answering the question as to what is the value of x.

The type of equation discussed above is sometimes designated as the interrogative sentence of algebra. By implication it asks for the value of the variable which will satisfy the equation. There is just one such value if the equation is linear; two if it is quadratic, and so on; but the values which the variable may take are strictly limited in number according to the degree of the equation.

Another and different type of interrogative equation is encountered as soon as two or more variables enter into the equation. The question now is no longer one of a single value, but of many values, and this is true regardless of the degree of the equation. Two cases must be distinguished, according as the values which the variables may take do or do not have to satisfy certain conditions.

The first case may be illustrated by the equation $x + y = 11$ in which either x or y may take any finite value whatever, subject, however, to the peculiar conditions that once a particular value has been assigned to either variable, then there is a limited number (in this case *one*) of values that the other variable may assume. In the case of the equation $x + 2 = 9$ we have seen (see p. 85) that the associative skill essential to an understanding of the statement of the equation consists in observing that the second member of the equation, whatever it may be, is 2 more than x. Comprehension of the meaning of the equation $x + y = 11$ necessitates as an associative skill recognition of the fact that the *sum* of two numbers equals something. The portion of the statement of the equation embraced in $x + y =$ conveys that idea. The particular further fact that attaches to this particular job is that the sum of the two numbers is 11, but the general, transferable, element that is common to this and every similar equation is the idea conveyed by the first member of the equation, and it is to be noted that this general idea is not in any manner affected by what happens to stand in the second member of the equation.

It is to be noted also that the expression $x + y =$ conveys the idea, not only that the second member of the equation is the sum of x and y, but that the second member is y more than x, and also x more than y. The meaning and processes of simultaneity are bound up with this latter idea. If

$$x + y = 11$$

and

$$2x + y = 13$$

at the same time the fact that in either instance y can be expressed in terms of a relation to x furnishes the most fundamental basis for the substitution of the value found for y in either of the original equations that is not used in expressing this value.

Of types of relationships between values that enter into an equation there remains one in which there is no restriction whatever imposed on one variable by any value which attaches to another. To this kind of equation the name identity is given. It is illustrated by

$$(a + b)^2 = a^2 + 2ab + b^2$$

or

$$x - y = \frac{x^2 - y^2}{x + y}$$

in which the values of the variables may take on any measure whatever of a real algebraic number. The peculiar property of this kind of equation is that while it states a fact it answers no questions. This form of the equation emphasizes to the highest degree manipulative skills, while associative skills of relation are practically lacking. The only reason for mentioning the identity in a chapter devoted to skills of relation is that the form of the identical expression is to all external appearances precisely like the equation of functional relationship which is entitled to some consideration here.

We have already discussed at length the capacity of the algebraic expression for the expression of a general number (see Associative Skills of Interpretation, page 59). This property of algebraic numbers also permits of the expression of general relations. If a represents the area of a square and s the length of the side, a general relation between area and side is expressed by $a = s^2$ but from a mathematical standpoint the important fact is not the mere statement—words would do as well as symbols for that; but the value of this type of expression is in translating the ideas of area and of side into the corresponding numerical measures of these magnitudes. The associative skill of this relationship takes account of the fact of the statement, but more particularly it perceives that a depends upon s for its value; that any change in s produces some change in a. In other words, a is a function of s.

A similar relationship exists between d, r, and t in the equation $d = rt$ and between F and C in

$$F = \frac{9}{5}C + 32.$$

The idea of function is always present in an equation containing two or more variables, but because the function concept pervades all of algebra wherever any general expression of symbolism of relation occurs, the function is classified and discussed under Skills of Interpretation, page 76.

When equations like the above are employed to express relationships that are found in situations which have actual and practical applications, the equations are called *formulas*. The formula thus derives its name and its identification as such not from its mathematical forms but from the use to which it is put. From a mathematical standpoint it is immaterial whether the word *equation* or *formula* is used, though it should be noted that the formula always involves the idea of function because of the presence of at least two variables.[2]

The Quadratic Equation

The quadratic equation, by definition, is of the second degree. Its solution is of the first degree. This suggests that between equations of the second degree and equations of the first degree there must be some relation. The discovery of that relation constitutes a method of solution. A relation, observable early, is that a quadratic expression results from the multiplication of two linear factors, and a quadratic equation may be obtained by multiplying, member by member (see axioms), two linear equations.

Often the detection of these original multipliers is simple, as in the case of $x^2 + 5x + 6 = 0$. Obviously this may be written $(x + 2)\ (x + 3) = 0$. Since 0 as the result of a product occurs only where one of the factors is 0, either $x + 2 = 0$ or $x + 3 = 0$, and these equations yield the results -2 or -3 as values which x may take in the original equation. These are the desired results, if, when substituted for x, the two members of the equation are made equal (in this case 0).

But some equations are difficult to factor. For instance $x^2 + 5x + 7 = 0$. The method of determining the factors is fairly simple but not altogether evident. It depends upon the formation of the only equation that is of such a nature that the factors which contribute to the product may surely and easily be detected in the product.

In the equation first discussed, namely, $x^2 + 5x + 6 = 0$, it was easy both to factor the first member and to determine what value the variable takes in each factor, because the second member of the equation is 0. Had the second member been any other number than 0, then it would not have been easy to determine what value to attach to the factors. Suppose that instead of the equation $x^2 + 5x + 6 = 0$ we had been given the equation $x^2 + 5x + 6 = 4$; the first member is still factorable, so that we could write $(x + 2)\ (x + 3) = 4$ but we should be unable to tell from this form what values to assign to x in order to make the product of the two factors equal to 4. Trial-and-error methods

[2] A suggestive list of exercises designed to illustrate functional relationship is to be found in Chapter X of E. L. Thorndike's *The Psychology of Algebra*. The Macmillan Co., 1924.

will be of no avail. It is doubtful if by this method a person would be able in a lifetime to find the right value of x to make the product of the two factors $x + 2$ and $x + 3$ equal to 4.

Had the two factors of the first member been equal, the solution would have been simple. Suppose we had been given such an equation as $(x + 3)\ (x + 3) = 4$ we could immediately have applied the axiom of multiplication by noting that if we have given an equation of the form $x + a = b$ the product of these two equal quantities $x + a$ and b by $x + a$ and b, respectively, yields the quantities $(x + a)\ (x + a)$ or $(x + a)^2$ and $b \cdot b$ or b^2, which, by the axiom of multiplication are equal; or, as usually written, $(x + a)^2 = b^2$.

It may be noted that had the original equation been $x + a = - b$, the process of multiplying each member by itself would also have been the same equation as we have derived above; namely, $(x + a)^2 = b^2$.

But the essential feature of the operation that is to be observed consists of the fact that if we have given a quadratic equation of the form $(x + a)^2 = b^2$ we may always know what factors produced the equation, by applying the axiom of multiplication. This fact supplies the associative skill for the solution of the quadratic.

Suppose the quadratic equation is of the form $x^2 + px + q = 0$. We now know that a convenient form for the equation to take is that of $(x + a)^2 = b^2$. If we can change the form of the given equation so as to represent two products of equal factors, the solution is made possible. To be able to do this depends upon the ability to recognize the relations of the terms of a perfect square of a binomial to each other and to the original binomial. This is effected by analyzing the square of any binomial. For instance, the square of $m + n$ or $(m + n)^2 = m^2 + 2mn + n^2$, or if it is desired to emphasize the fact that the first term is a square of m and that the second term also contains m, the equation may be written in the form of $(m + n)^2 = m^2 + 2n \cdot m + n^2$. The third term now appears to be the square of half the coefficient of m, where $2n$ is considered as a coefficient. Since our original binomial was *any* binomial, the conclusion just reached regarding the third term must apply generally to the relations between the three terms of a product of a binomial by itself.

Going back now to the equation $x^2 + px + q = 0$, which is to be solved for x, it is observed that the square of the two terms in x may be completed as above by adding to them the square of half the coefficient of x, or the square of $\frac{p}{2}$, which equals $\frac{p^2}{4}$.

The expression $x^2 + px + \frac{p^2}{4}$, then, is a perfect square. If q in the equation is already equal to $\frac{p^2}{4}$ nothing further need be done to complete the square, but if q is not equal to $\frac{p^2}{4}$ then there must be added to q the difference between $\frac{p^2}{4}$ and q which is $\frac{p^2}{4} - q$. Note that if $\frac{p^2}{4} - q$ is added to the first member of the equation the sum is $x^2 + px + \frac{p^2}{4}$.

The meaning of an equation necessitates adding the same quantity also to the second member so that the transformed equation becomes

$$x^2 + px + \frac{p^2}{4} = \frac{p^2}{4} - q$$

or

$$\left(x + \frac{p}{2}\right)^2 = \frac{p^2}{4} - q.$$

The two factors of each member are now seen to be equal, if the square root of $\frac{p^2}{4} - q$ is considered. The application of the axiom of multiplication then enables us to see that originally the linear equation from which the quadratic equation might have been considered as derived was

$$(1) \quad x + \frac{p}{2} = \left(\frac{p^2}{4} - q\right)^{\frac{1}{2}}$$

or

$$(2) \quad x + \frac{p}{2} = -\left(\frac{p^2}{4} - q\right)^{\frac{1}{2}}$$

from which the factors derived from the two equations become

$$x + \frac{p}{2} - \left(\frac{p^2}{4} - q\right)^{\frac{1}{2}}$$

and

$$x + \frac{p}{2} + \left(\frac{p^2}{4} - q\right)^{\frac{1}{2}}$$

Or, the original equation might have been written

$$\left[x + \frac{p}{2} - \left(\frac{p^2}{4} - q\right)^{\frac{1}{2}}\right]\left[x + \frac{p}{2} + \left(\frac{p^2}{4} - q\right)^{\frac{1}{2}}\right] = 0.$$

Summary of Associative Skills of the Quadratic Equation

1. A quadratic equation may always be considered as obtained from two linear equations by a process of multiplication.

2. The linear equations which produce a quadratic equation by multiplication may always be determined if the quadratic equation is of the form $(x + a)^2 = b^2$.

3. The third term of a result obtained by multiplying a binomial by itself may be obtained from the other two terms.

Associative Skills Illustrated in the Interpretation of a Formula

A stated formula has two principal algebraic applications: (1) the derivation, by substitution of numerical values, of some part (unknown) of the formula; (2) changing the subject of the formula.

In the formula $T = 2\pi rh + 2\pi r^2$ the derivation of T, having given the values of π, r, and h, is a very simple matter for one who possesses the requisite ability in reading, or, as referred to here, the necessary skills of interpretation. Assume that the numerical values of π, r, and h are $3\frac{1}{7}$, 4, and 20, respectively. The following implied questions must be answered, in order to determine the value of T:

What is the meaning of $2\pi rh$? That is, what operation is understood between 2, π, r, and h? (See page 62 and page 71.)

How then shall the value of the term $2\pi rh$ be found in this particular exercise?

What is the meaning of r^2 in the last term? (See page 62.)

In this case what is to be done to obtain the value of $2\pi r^2$?

May we now state the value of T? [Not until the values that we have found for $2\pi rh$ and for $2\pi r^2$ have been added.] (See page 65.)

Suppose that instead of desiring the value of T, having given the numerical values of π, r, and h, it is desired to change the form of the formula so as to obtain an expression for h. This is called by Nunn "changing the subject of the formula." Now it is necessary to think, not in terms of particular values of particular symbols and the way in which they are combined, but in terms of the manner in which h is related to the entire equation. The subject of the formula is still T, but the subject of our thinking becomes h, which is finally to become the subject of the changed formula.

Now the question is, What does the first member equal in values of either of the terms of the second member? (See page 89.)

T is more than $2\pi rh$ by the value of $2\pi r^2$, or T is more than $2\pi r^2$ by the value of $2\pi rh$. We shall use the relation that seems most desirable. (See page 88.)

Since T is more than $2\pi rh$ by the value of $2\pi r^2$, it follows that $T - 2\pi r^2 = 2\pi rh$.

It now appears that $T - 2\pi r^2$ is how related to h? ($T - 2\pi r^2$ equals h, *multiplied by* $2\pi r$.)

Then h must equal $T - 2\pi r^2$ divided by $2\pi r$.

Our final conclusion is $h = \dfrac{T - 2\pi r^2}{2\pi r}$.

Can this expression in the second member be further simplified? (See page 61 and page 64.)

CHAPTER VIII

GENERAL SUMMARY

Algebra is distinguished for the variety of the tasks that can be set, in the performance of which there is required the exercise of some special ability. To each one there may be applied a separate manipulative skill. So far as the problems and exercises themselves are concerned, it is not unlikely that a machine could be devised that could solve each of them; a great many machines would be necessary, and still this would not be algebra. It is not difficult to classify work in algebra according to the length or character of the symbolism employed. For instance, all topics could be grouped according to the number of terms employed, whether monominal, binominal, or trinominal, and so on; or according to the number of operations involved; or according to whether the numerals used were integral or fractional; and in many other ways.

Each type could easily be subdivided until we should have finally each task placed with reference to its particular technique. When we had done this all that would be necessary would be to identify the type to which a task might belong, and proceed to set in motion the appropriate machinery, mental or otherwise, for its treatment. Logically this plan appears quite perfect. In practice it proves to have at least three fatal defects. In the first place the number of separate jobs is too great to be mastered in this way; in the second place, ability gained in one situation does not easily transfer to another, which illustrates the same principle but differs in the external appearance; and in the third place, ability to manipulate seems to cultivate but slight sense of awareness of applicability of the skill which the manipulation employs. Instances of the presence of such defects are cited in the chapters on "The Meaning of Algebra," "The Nature of Skills," and "Learning in Algebra."

On the other hand, the manipulative instruments, as distinguished from tasks, by which the operations of algebra are effected, are few in number and the principles employed are also restricted to a narrow field. Elementary algebra requires that the pupil shall be able to add, subtract, multiply, and divide algebraic numbers either as found in independent expressions or in equations. While the range of the kind of work that is performed is very limited, it is easy to create an impression in the mind of the pupil of the existence of a great number of different operations. It is evident, for instance, that pupils turn with very little awareness of similar elements from practice in solving equations of the type:

Solve for x:

(1) $4x + 5 = 17$

to problems of the type:

Solve for T:

(2) $A = P + Prt.$

In these instances the pupils who were able to solve for x in the first exercise with 95.7 per cent accuracy, were disturbed by the second to such an extent that only 64.2 per cent solved it.[1] These were the results obtained in spite of the fact that practice preceded each of the tests, that between the giving of the tests on the first and second problems the pupils had a great deal of experience with equations, and the further fact that the algebraic principles involved in the two exercises are identically the same. The first of these problems was attempted early in the ninth grade; the second toward the close of the year.

On the surface the statistical evidence on the outcome in the two instances indicates that the first exercise was well taught and that the second was poorly taught. If the exercises are to be considered as separate jobs, the conclusions stated are fair ones. If algebra is to be considered as a subject designed to inculcate principles and stimulate both a method of, and an ability in, thinking, then the statements that one problem was well taught and that the other was poorly taught are inconsistent.

From the pupils' viewpoint there was evidently something quite different about the exercises. To the pupil who attempts the solution of these two equations on a manipulative basis the difference in symbolism is enough to account for an appearance

[1] W. D. Reeve, *A Diagnostic Study of the Teaching Problems in High School Mathematics*, pp. 104, 109. Ginn and Co., 1926.

of a totally new set of circumstances as he passes from the first to the second of the equations.

To the pupil schooled in the associative skills of appreciation, and relation suggested by the equations, the problems become what they really are, exercises in the same algebraic principles, with changed numbers but not with changed methods of procedure or changed meanings of any kind.

The following simple exercises are taken to illustrate the manner in which the associative skills of interpretation and relation may be applied:

Problem	*Problem*
Solve for x: $4x + 5 = 17$	Solve for t: $A = P + Prt$
Skills of Interpretation	
$4x + 5$ is a number which equals the number 17.	A is a number which equals the number $P + Prt$.
$4x$ is a number, as are also 4 and x.	Prt is a number as are also P, r, t, Pr, and Pt.
Skills of Relation	
x is a number whose relation to the rest of the equation is sought:	t is the number whose relation to the rest of the equation is sought:
17 is 5 more than $4x$:	A is P more than Prt:
therefore $4x = 17 - 5 = 12$.	therefore $A - P = Prt$.
12 is 4 times x:	$A - P$ is Pr times t;
therefore $x = \frac{12}{4} = 3$.	therefore $t = \frac{A - P}{Pr}$.

It will be observed that the treatment given these exercises differs from the usual appeal to "axiomatic" methods in that the use of the associative skills keeps constantly to the forefront of attention the actual relations between the numbers which are involved in the equations. Throughout the entire process the emphasis is placed not on the performance of some operation to be selected from "a bag of tricks" but upon the relationships which depend for their further simplification only upon the most elementary notions of quantities. By elementary notions is meant those of *more than* and *less than*. If a child in transferring a large number of books from the study room to the library were to attempt to carry more than he could lift, and in such a circumstance were undecided whether to increase the burden by placing more books on the load, in the presence of *more than*, or to lighten the load by taking some off, it is extremely improbable that associative skills would make much of an appeal to his intelligence. Not only must a child have sufficient mental ability to make the correct response in the presence of the situation itself, but he must possess enough imagination to comprehend the setting and make a decision when all of the actual circumstances are remote and there is stimulation of the senses only through whatever representation may be conveyed by symbols.

To answer the question, "What is x, if 12 is 5 more than x?" is of tremendous difficulty compared with satisfying the demand, "Subtract 5 from 12," although both require the same manipulative skill. In the one case judgment of a high order is required. In the other case, a mere act, in response to a definite stimulus, suffices.

In this study there have been discussed at some length forty-four associative skills of ninth grade algebra. The summaries to which the index below refers are necessarily, in most cases, so condensed as to afford only a name for the skills, whose detailed characteristics are outlined in the pages preceding the summaries.

Outline of Associative Skills of Ninth Grade Algebra

No. of Skills	Character	Summarized on Page No.
	Directed Numbers	
6	General Meaning	41-42
2	Addition	46
1	Subtraction	53
4	Multiplication	55-56
1	Division	58
	Algebraic Interpretation	
14	Fundamental Operations	73-75
3	General	77
2	Line Graph	83
	Equations	
7	Linear	88-89
1	Systems	90
3	Quadratic	95

The associative skills that have been listed center attention on the thinking that may be a part of algebraic operations. They are not devices, nor do they encourage habits that will have to be broken later or methods of thinking that are calculated to hinder a pupil as power develops in dealing with more complicated algebraic conceptions.

The pupil who initially acquires the power to solve the equation $\frac{3x}{4} = 7$ because he has learned to read and interpret the equation as meaning that the second number of the equation as it stands is one-fourth of $3x$, followed by the idea that the second number of the resulting equation is 3 times x, has been called upon to exercise the elementary ideas pertaining to the number of fourths in a whole and the relation of 3 times a quantity to the quantity itself. Relations are stressed from the first, but there is nothing in this approach that hinders the pupil from generalizing to the effect that if equals are multiplied by equals the results are equal, or that if equals are divided by equals the results are equal. These axioms have an opportunity to grow out of the pupil's experience and to take their proper place in his estimation as an economical form of thought.

Contrast the approach which has just been discussed with the situation which confronts a pupil when he meets, for the first time, an equation of the form $\frac{3x}{4} = 7$ and is asked to deal with it by applying the axioms. Aside from the fact that this presents a problem that is new both in the form of its language and in the treatment of its solution, the "axiomatic" approach has the fundamental defect of immediately centering attention upon manipulation; although manipulation has its place, and a valuable place, in algebra, it has been shown by specific illustrations that mere ability to perform algebraic operations is very small indication either that the application of a pure manipulative ability will be made at the right time or that the possibility for its use will be recognized later.

The advantages of emphasizing these associative skills are threefold: They are not so numerous as the manipulative skills, but they give purpose and direction to these skills. The associative skills, at every step, put meanings into operations by inquiring into the significance of the relations between the number ideas that compose algebraic situations. And, finally, the associative skills provide the pupil, at every step, with an opportunity to check upon his own conclusions.

All of the elementary exercises that a pupil meets, for instance, that involve the idea of sign are directly related to the number scale. In place of the multiplicity of rules governing addition, subtraction, multiplication, and division, the associative skills of directed number necessitate the fundamental concept of the meaning of algebraic number and direct practice around this central idea.

In dealing with equations fundamental notions of number relations are given an opportunity to operate until formal methods and ideas may develop as final forms of economy of thought, but such methods have the advantage of a background of appreciation of significant relations, which the operation of rules does not seem to inculcate.

In so simple an equation as $\frac{x}{3} = 7$ the difference between realizing that "the second member of the equation is one-third of x," and proceeding to a solution by observing the rule of "multiply both numbers of the equation by 3" is the difference between a process that calls for the exercise of the pupil's own thought to reach a conclusion, and a manipulative act into which there may have entered little thinking.

APPENDIX

Section I

The following are exercises whose treatment includes the skills of first-year algebra.

Exercise	[1] Per Cent Right
1. Perform the indicated operations and reduce to simplest form: $$\frac{2\sqrt{5c}\ .\ \sqrt{12c}}{3c\sqrt{15}}$$	9.6
2. Perform the indicated operations and simplify the results in the following: $$\frac{1}{a-b}-\frac{3a}{a^2-b^2}$$	15.1
3. Perform the indicated operations and simplify the result in the following: $$5-\frac{2-3a}{5}-4a=$$	15.8
4. Solve for x: $\frac{5x}{2}-1=\frac{17x}{3}-\frac{5}{3}-2\frac{1}{2}x$	16.2
5. Perform the indicated operations and simplify the result in the following: $$\frac{x-y}{x^2+y^2}\cdot\frac{x^4-y^4}{(x-y)^2}=$$	16.9
6. Perform the indicated operations and simplify the result: $$\frac{8x+5y}{3xy^2}-\frac{2x+3y}{x^2y}$$	17.4
7. Solve for r: $$S=\frac{a}{1-r}$$	19.0
8. Find the value of $$\frac{3x(x^2-5xy+3y^2)}{x-3y}$$ for $x=2$ and $y=-3$	21.4
9. Solve for x: $$x^2+x-3=0$$	21.4
10. $(2x^3my^2)^2=$	22
11. Solve for x: $$\frac{x+4}{x-5}=\frac{x+2}{x-3}$$	22.3

[1] As determined by tests of approximately 1,200 pupils.

Exercise	Per Cent Right[1]
12. Perform the following operation and simplify the result: $$\frac{7a+2}{12a}+\frac{4a-3}{4a}=$$	22.7
13. Solve for x: $$\frac{5}{2-x}=\frac{3}{4+x}$$	27.2
14. Perform the indicated operations and simplify the result: $$2x+2(3x-2y)-4(x+3y)=$$	28.7
15. Perform the indicated operations: $$4x-6y+5x-3y=$$	29.3
16. The volume of a sphere is the cube of its radius multiplied by $\frac{4}{3}\pi$. Find V if $r=12$ ft.	29.3
17. $2a^{1/2}\times 5a^{3/2}=$	29.3
18. Find the prime factors of: $$2x^2z+11xz+12z=$$	29.9
19. Find the prime factors of: $$(a+b)^2-9$$	32.1
20. Solve for x: $$\frac{5x}{6}+\frac{25}{12}=3+\frac{x}{4}$$	36.5
21. Find the product: $$(3br)\ (4b^2r)\ (6b^3r)=$$	36.8
22. Solve for r: $$C=\frac{E}{R+r}$$	36.9
23. Find the prime factors of: $$x^4-y^4$$	37.4
24. Perform the indicated operations and simplify the result in the following: $$\frac{a}{x^2-y^2}=\frac{x+y}{b}$$	38.6
25. Solve for x: $$x-3\ (2-5x)=32$$	38.9
26. Divide as indicated: $$\frac{8x}{12}\div\frac{9x^3}{15}=$$	38.9
27. Solve for y: $$8y-14-3y+9=0$$	39.6
28. Solve for M: $$L=\frac{Mx-y}{x}$$	40.0
29. Perform the indicated operations and simplify the result in the following: $$7x-[5y-(4x-3z)+(7y-6z)]$$	40.1

[1] As determined by tests of approximately 1,200 pupils.

Exercise	Per Cent Right[1]
30. Solve for x: $$x^2 + 4x = 18$$	41.3
31. Select three pairs of corresponding values for x and y in the following equation and draw its graph: $$2x - y = 6$$	41.4
32. Graph the system of equations below and indicate whether there is a solution: $$4x + 3y = 2$$ $$8x + 6y = 10$$	42.6
33. Simplify: $$\sqrt{12} + \sqrt{48} + \sqrt{\frac{1}{3}} =$$	44.2
34. Solve for r: $$v = \frac{4}{3} \pi r^3$$	44.3
35. Solve for x: $$x^2 - 23x = -112$$	44.6
36. Solve for x: $$\frac{x}{2} = \frac{7}{4} - \frac{x}{3}$$	44.7
37. Solve for x: $$\frac{5(x+2)}{6} + x = \frac{x}{2} + 5$$	44.8
38. Solve for x: $$0.4x - 5 = 3.8$$	45.3
39. Find the prime factors of: $$4a^2 + 16a + 16$$	45.6
40. Divide as indicated: $$\frac{-15x^3y^3(7-3)}{-5x^2y^2}$$	46.0
41. Solve for x and y: $$\frac{x}{5} + \frac{y}{2} = 4$$ $$\frac{x}{3} = \frac{y}{2} - 12$$	46.4
42. Perform the indicated operations and simplify the results in the following: $$\frac{2a}{3} + \frac{5a}{8} - \frac{a}{4} =$$	46.5
43. Solve for x: $$12 - 8x = 3 - 2x$$	46.9
44. Find the prime factors of: $$3x^2 + 6x$$	47.4
45. Graph the following system of equations and indicate whether it can be solved: $$x + y = 6$$ $$2x - 3y = 2$$	48.4

[1] As determined by tests of approximately 1,200 pupils.

Section II

The table and other data appearing below pertain to five typical exercises attempted by 549 candidates for the examination of the College Entrance Examination Board shortly prior to the date of the 1927 examination:

	PROBLEM NO.				
	1	2	3	4	5
Number of pupils..........	549	549	549	549	549
Number of exercises correct.	489	523	487	452	405
Number omitted	1	11	0	5	59
Number wrong	59	15	62	92	85
Per cent correct...........	89.0	95.2	88.7	82.3	73.7

Problem 1

I find that $\frac{16a^3 - 8a^5}{4a^2} = 4a - 2a^3$ or $2a(2 - a^2)$.

Incorrect results:

(1) $2a(2 - a)$
(2) 2
(3) $2a$
(4) $4a - 2a^2$
(5) $4a(1 - 2a^4)$
(6) $4a - 8a^5$
(7) $2a(-a^2)$
(8) $2a(2 + a)(2 - a)$
(9) $2(2a - a^3)$
(10) $4a^2 - 2a^3$
(11) $\frac{4a - 2a^3}{a^2}$
(12) $a(2 - a^2)$
(13) $2a(1 + a)(1 - a)$
(14) $[-2a(a^2 - 2)]$
(15) $a(1 - 2a^2)$
(16) $4a^2(4a - 2a^3)$
(17) $\frac{2}{a^4}$
(18) $2a(2 - a^3)$
(19) $-4a^3$
(20) $4a - a^3$
(21) $2a^3(2 - a^2)$
(22) $2(2 - a^2)$
(23) $-2a(a - 2)(a + 2)$
(24) $a(1 - a^2)$
(25) $8a^3(2 - a^2)$
(26) $14a^3$
(27) $4 - 2a^3$
(28) $4a(1 - 8a^4)$

Problem 2

Find the sides of a triangle if the second side is 4 ft. longer than the first, the third side 6 ft. longer than the first, and the perimeter 40 ft. The equation which expresses the condition of this problem is $3x + 10 = 40$ or $x + x + 4 + x + 6 = 40$.

Incorrect results:

(1) $P = 2(a + b)$
(2) $3x + 10 = 24$
(3) $6x + 20 = 40$
(4) $10 + 14 + 16 = 40$
(5) $x + 4x + 6x = 40$
(6) $3x + 6 = 40$
(7) $\frac{x + 6}{2} = 40$
(8) $3x + 14 = 40$
(9) $3x + 10 = 30$
(10) $3x = 50$
(11) $40 = 6x + 20$

Problem 3

The result of subtracting $-5a+7b-75$ from $+8a-12b-156$ is $13a-19b-81$.

Incorrect results:

(1) $3a-5b-131$
(2) $3a-5b+231$
(3) $13a-19b-79$
(4) $13a-19b+81$
(5) $3a-5b-231$
(6) $3a-29b-81$
(7) $13a-29b-81$
(8) $13a+19b-81$
(9) $13a-19b-231$
(10) $13a-19b-71$
(11) $11a-19b-81$
(12) $-13a+19b+81$
(13) $13a-19b-79$
(14) $3a-5b-23$
(15) $15a-19b-81$
(16) $-3a-5b-81$
(17) $-13a+19b-81$
(18) $3a-5b-231$
(19) $13a-15b-81$
(20) $13a+5b-81$
(21) $13a-19b-75$
(22) $3a-19b-81$
(23) $3a-19b-231$
(24) $13a+9b-141$
(25) $6a-19b+81$
(26) $14a-19b-81$

Problem 4

In solving the formula $A=\frac{1}{2}h(b+b')$ for h, I find that $h=\frac{2A}{b+b'}$.

Incorrect results:

(1) $\frac{a}{b}$
(2) $\frac{a}{\frac{b+b'}{2}}$
(3) $\frac{2b}{2a}$
(4) $\frac{2a}{2b+2b'}$
(5) $\frac{b+b'}{2A}$
(6) $\frac{a}{2b}$
(7) $\frac{2a}{2b}$
(8) $\frac{a}{2(b+b')}$
(9) $\frac{a}{b+b'}$
(10) $\frac{2a-b'h}{b}$
(11) $\frac{h(b+b')}{2}$
(12) $A\left(\frac{4}{b}\right)$
(13) $2A'$
(14) $\frac{A}{2(b+b')}$
(15) $\frac{2a}{\frac{(b+b')}{2}}$
(16) $\frac{A}{4b}$
(17) $\frac{2a-hb'}{2}$
(18) $\frac{2a-b}{b}$
(19) $\frac{A}{(\frac{1}{2}b+\frac{1}{2}b')}$
(20) $h=\frac{1}{2}a(b+b')$
(21) $\frac{1}{2}\,\frac{A}{b+b'}$
(22) $\frac{2A-hb'}{b}$
(23) $\frac{2b+2b'}{A}$
(24) $\frac{4a}{b+b'}$
(25) $\frac{\frac{1}{2}b+\frac{1}{2}b'}{A}$
(26) $\frac{2A}{b}$

Problem 5

A ladder 20 ft. long reaches a window 16 ft. above the ground. Assuming the ground is level, how far is the foot of the ladder from the foot of the wall? The equation which expresses the conditions of this problem is $20^2 = 16^2 + x^2$ or $400 = 256 + x^2$.

Incorrect results:

(1) $\frac{16}{20} = \frac{x}{16}$

(2) $\frac{16}{20} = \text{Sin } A$

(3) $\frac{16}{24} = \frac{x}{16}$

(4) $\frac{B}{16} = \text{Cos } B\frac{20}{16}$

(5) $x = \frac{16}{20}$

(6) $\text{Cos } x = \frac{16}{20}$

(7) $x = 20 - 16$

(8) $\text{Sin } \Delta\ B = \frac{16}{20}$

(9) $\text{Sin } 72° = \frac{x}{20}$

(10) $x = 144$

(11) $\text{Sin } A = \frac{G}{C}$

(12) $\text{Log tan } \angle A = \frac{\log 16}{\log x}$

(13) $\frac{16}{x} = \frac{16}{20}$

(14) $x = \frac{20}{16}$

(15) $20 : x = x : 16$

(16) $\text{Tan } x = \frac{24}{16} = \frac{5}{4}$

(17) $\frac{x}{20} = \frac{16}{20}$

(18) $\frac{\tan}{a} = \frac{16}{x}$

(19) $x + 36 = 180$

(20) $\cos 90° = \frac{x}{24}$

(21) $\cos 90° = \frac{16}{20}$

(22) $\sin x = \frac{16}{20}$

(23) $\sin 36°\ 52' = \frac{a}{20}$

(24) $\frac{16}{x} = \frac{20}{16}$

(25) $400 = 176 + x^2$

(26) $\text{Sin } B = \frac{x}{20}$

(27) $\log \tan \angle\ A = \frac{a}{b} = 12$

(28) $\sqrt{20} = \sqrt{16^2 + x^2}$

(29) $400 = b^2 + 156$

(30) $\log \sin x = \log 16 - \log 20$

(31) $\tan \angle a = \frac{16}{x}$

(32) $x = \text{anti log } \{\log 16 - \log \tan [\text{anti log sin } (\log 16 - \log 20)]\}$

(33) $\text{Sin } A = \frac{a}{c}$

(34) $\text{Cos } a = \frac{x}{20}$

(35) $AC = \frac{16}{\tan [5307}$

(36) $\text{Log tan } A = \log 20 - \log x$

(37) $\tan B = \frac{16}{20}$

(38) $\log \sin x = 1.2041 - 1.3010$

(39) $\frac{16}{20} = \frac{20}{x}$

(40) $\sin x = 16$

(41) $\sin A = \frac{\text{opposite}}{\text{hypotenuse}}$

(42) $16 = 20 - x$

(43) $\frac{16}{x} = \frac{x}{20}$

(44) $\text{Sin } A = \frac{a}{b}$

(45) $\text{Cos } 53°\ 8' = \frac{6}{20}$

Problem 5 (Continued)

Incorrect results:

(46) $\frac{4}{5} = \frac{16}{y}$

(47) $\sin x = \frac{x}{20}$

(48) $16x = 20$ (or) 400

(49) $20 - 16 = 4$

(50) $\cos 74° = \frac{b}{20}$

(51) $x^2 = 300,\ x = 10\sqrt{3}$

(52) $h^2 - a^2 = b^2$

(53) $A = \frac{a}{c}$